AND YOU WILL HAVE BENEFITED FROM THE WORK OF UK ACADEMICS.

Written by: Dr Lee Elliot Major
Initial research: Dr Jenny Gristock
Universities UK project team: Dr Tony Bruce,
Alex Bols, Sam Hall, Simon Wright
Picture research: Zooid Pictures Limited

ISBN 1 84036 129 8
June 2006

Universities UK
Woburn House
20 Tavistock Square
London, WC1H 9HQ

www.UniversitiesUK.ac.uk

Designed by: Nick Bell Design
Printed by: Zwaan Printmedia, The Netherlands

EurekaUK

100 DISCOVERIES AND DEVELOPMENTS IN UK UNIVERSITIES THAT HAVE CHANGED THE WORLD

PLANES, TRAINS AND AUTOMOBILES

IDEAS FOR IDEALS

UNDERSTANDING OURSELVES

UNDERSTANDING OUR ENVIRONMENT

SPACE EXPLORATION

INTRODUCTION

Academics are still popularly perceived as brilliant but eccentric, impractical minds preoccupied with ideas that no one in the 'real world' will understand, let alone be affected by. Nothing, however, could be further from the truth.

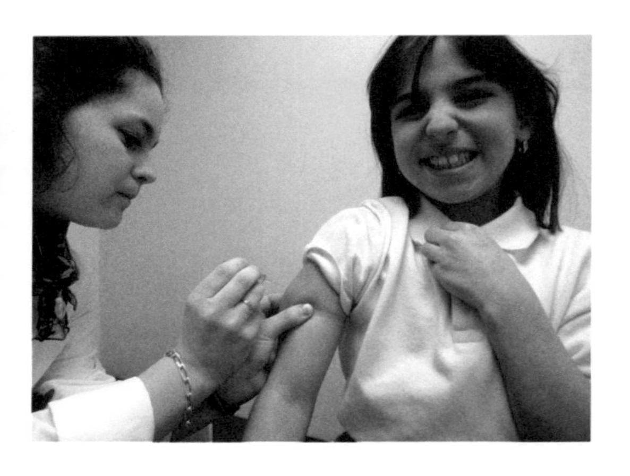

UK researchers in fact have transformed our world: allowing us to live healthier, safer and longer lives; providing globe-conquering, super-fast travel and communication; deepening our understanding of the world around us, and indeed of our own behaviour in the past and present.

Take a pill, undergo an operation, have a baby, go to school, receive a benefit, surf the Internet, take a journey by plane, train or automobile, play a CD, phone overseas, hear a weather forecast, follow a road sign, give up smoking, study the stars – and you will have benefited from the work of UK academics, usually without even knowing it.

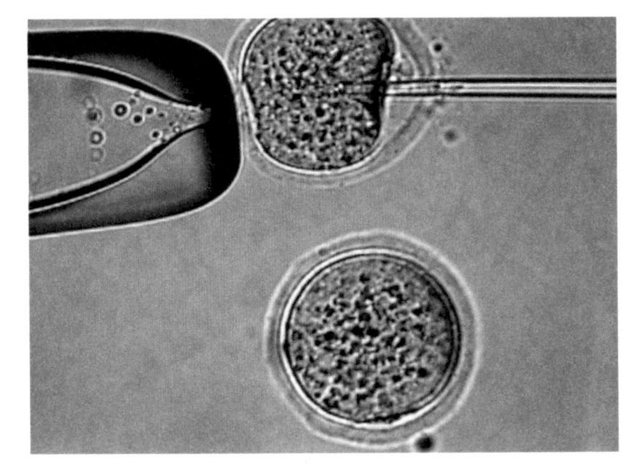

This book, as far as we know, is a first: it describes 100 major discoveries, developments and inventions made in UK universities during the last 50 years that have impacted on the world. The list is the result of a careful selection process. First, nominations were requested from individual universities; then a committee of university heads sifted through the proposed examples; and finally, claims were checked with academics and experts in specialist fields. (Long gone are the days of the polymath aware of developments across many academic disciplines.)

The aim of this list, however, is not to provide an exhaustive overview of the outputs of UK research. It is intended rather to provide a selection of case studies – real tangible examples of research – that reveal something about the nature of discovery and how it impacts on the world. The time period in which the discoveries were made (1953 to 2003) is also an arbitrary one. But we hope this is just the start of an on-going exercise.

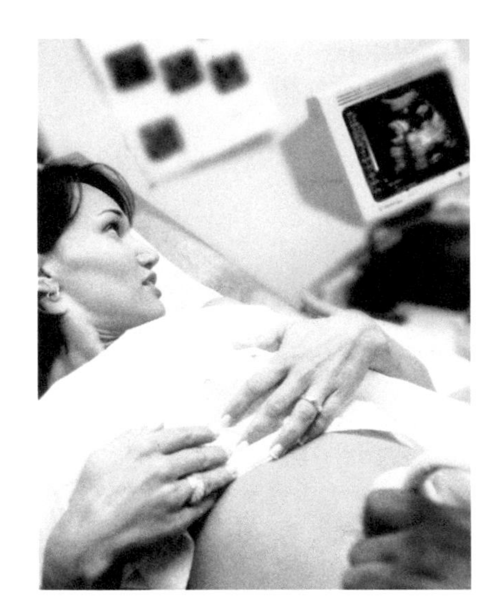

We defined an 'impact' as a discovery, development or invention that has had a major impact on people's everyday lives, or has improved our understanding in a field to such a degree that this change is comprehensible to everyone.

There are 83 entries (covering 100 discoveries or developments) grouped into nine sections, spanning the medical, physical, social sciences and the arts and humanities. Each chapter also includes an interview with, or profile of, one of the academics featured in the section.

The examples represent a tiny sub-set of the efforts of UK researchers, yet they alone suggest an incredible return on the money invested in universities and an almost immeasurable impact. Who can imagine a world now without the Contraceptive Pill, test-tube babies, or computers? The things we take for granted today offer a salutary tale for those in charge of research budgets: the proportion of Government spending on research in the UK still lags behind that of all other major industrialised nations.

Yet one of the recurring conclusions that emerges from the compilation is that research is by its nature unpredictable. Those eureka moments, or flashes of insight, can come at any moment – in a traffic jam, during a coffee break, in the bath. It is impossible to guess where or when or in which subject the next important discovery will emerge.

Ask academics, and many will also have no inkling of what future applications will be made possible by their research. Who would have guessed that the hunt for stellar black holes would lead to medical scans for cancer victims for example, or that an incidental observation by a psychiatrist would uncover what makes schools effective?

It should be stressed however that research results seldom offer complete solutions to life's problems. Our pool of knowledge is constantly being replenished, with previous findings rebutted or superceded. Debates will continue to swing back and forth on the received wisdom concerning for example, the sleeping position of babies, the cause of environmental problems, or the factors behind the patterns of history.

The unpredictable nature of research suggests that it should continue to be supported across a broad range of subjects. If strategic direction of research is required, it should apply at the broadest level. Ultimately, we have to trust academics on the cutting edge of research, not eminent committees, to identify the next exciting developments.

The case studies also demonstrate how long it takes (probably about 50 years) to measure the success of basic research, at least in terms of impacts on the wider world. This is why older universities dominate the examples cited in the publication. New universities have only in the last decade attracted (limited) research funds. Examples from these institutions are less advanced in terms of their impact on the world, for this reason.

Other common themes recur throughout the stories. Many academics featured are single-minded, stubborn characters. Often they succeed against the odds, working in unfavoured subjects with little Government support. Time and again, the spark that ignites a new development emerges when different disciplines come together. And in many cases, the partnerships involve an experienced academic working with a young research student.

Finally, it is the thrill of intellectual inquiry, not the lure of money or power, that drives these men and women. It is my hope that this talent will increasingly be matched by a similar adventurous spirit among companies in the UK, allowing entrepreneurs to reap the rewards (worth 100s of billions of dollars each year) produced by the country's rich harvest of discovery.

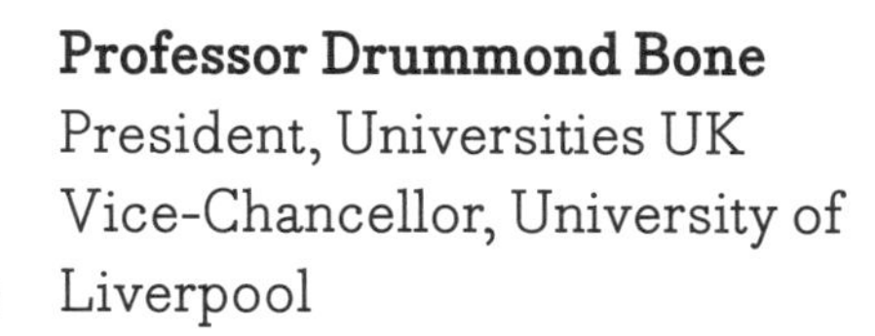

Professor Drummond Bone
President, Universities UK
Vice-Chancellor, University of Liverpool

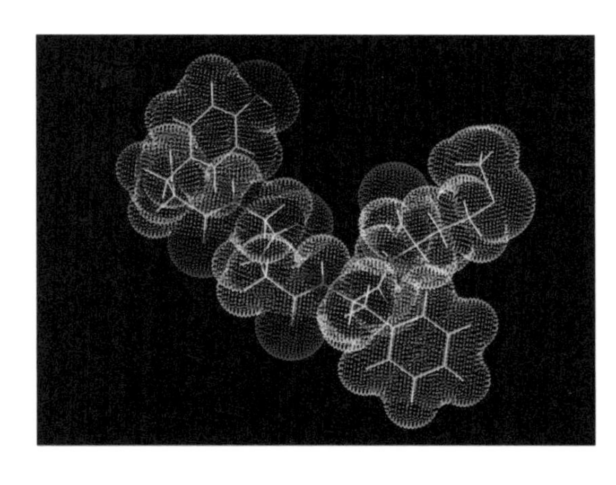

HEALTHY BABIES AND BIRTH CONTROL

Having a baby is still an unpredictable business. Any mother who has been rushed into hospital to give birth will know this only too well. Humans are particularly vulnerable when it comes to producing offspring: other animals are less likely to miscarry, or contract harmful diseases, and are generally more fertile.

But advances by UK medical researchers during the last 50 years have transformed pregnancy across the world into a safer, more controlled experience.

Who can imagine life now without the Contraceptive Pill? Developed by Herchel Smith in the 1960s the mass-produced Pill allowed women for the first time to be nearly 100% sure they would not fall pregnant after having sex. The development helped to empower women, allowing them much more control over when to have children.

Test-tube babies and modern in-vitro fertilisation (IVF) treatment have allowed infertile couples to have babies, when before this was simply unimaginable. One in five couples using IVF are now successful.

UK academics meanwhile have also developed the scans and tests that are now standard practice for doctors identifying problems and harmful diseases in babies.

Meanwhile advice for parents not to smoke during pregnancy – and to make sure babies sleep on their backs – has helped more babies to grow up to live longer, healthier lives.

Mass produced contraceptive tablets came on to the market in the 1960s, providing a safer, more accessible and affordable way of preventing pregnancies.

PRODUCING CHOICE

While birth control existed before, the Pill empowered women to make their own choices about whether or not they wanted children.

All this is due, in part, to the work of Herchel Smith, a researcher at the University of Manchester, who in 1961 developed an inexpensive way of producing chemicals that can stop women ovulating during their monthly menstrual cycle.

Naturally secreted by glands in our bodies, steroids are among the key chemical hormones that regulate the activities of our organs and bodies. Some steroids such as estrogen and progesterone can prevent a woman from releasing the eggs that are fertilised during pregnancy.

The hormones in early birth control pills were extracted from natural sources, usually the Mexican vegetable the yam. Previous methods for making steroid hormones were expensive and inefficient. Smith developed chemical reactions that produced the hormones in just the desired form, and went on to show how these hormones could be mass produced as the Contraceptive Pill.

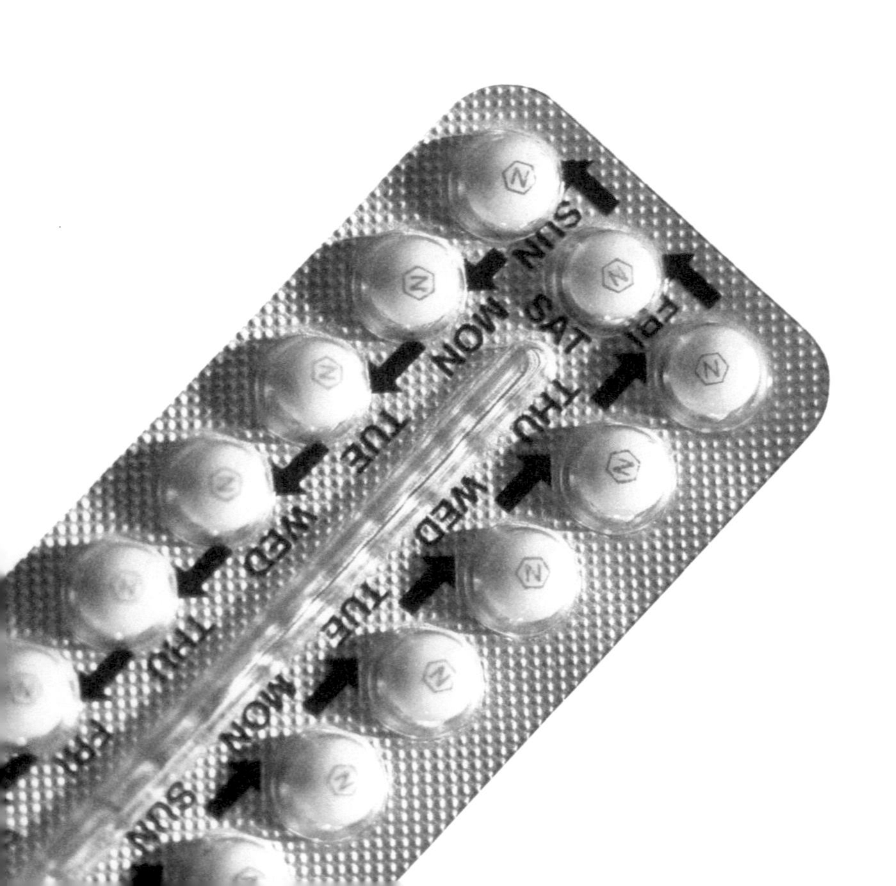

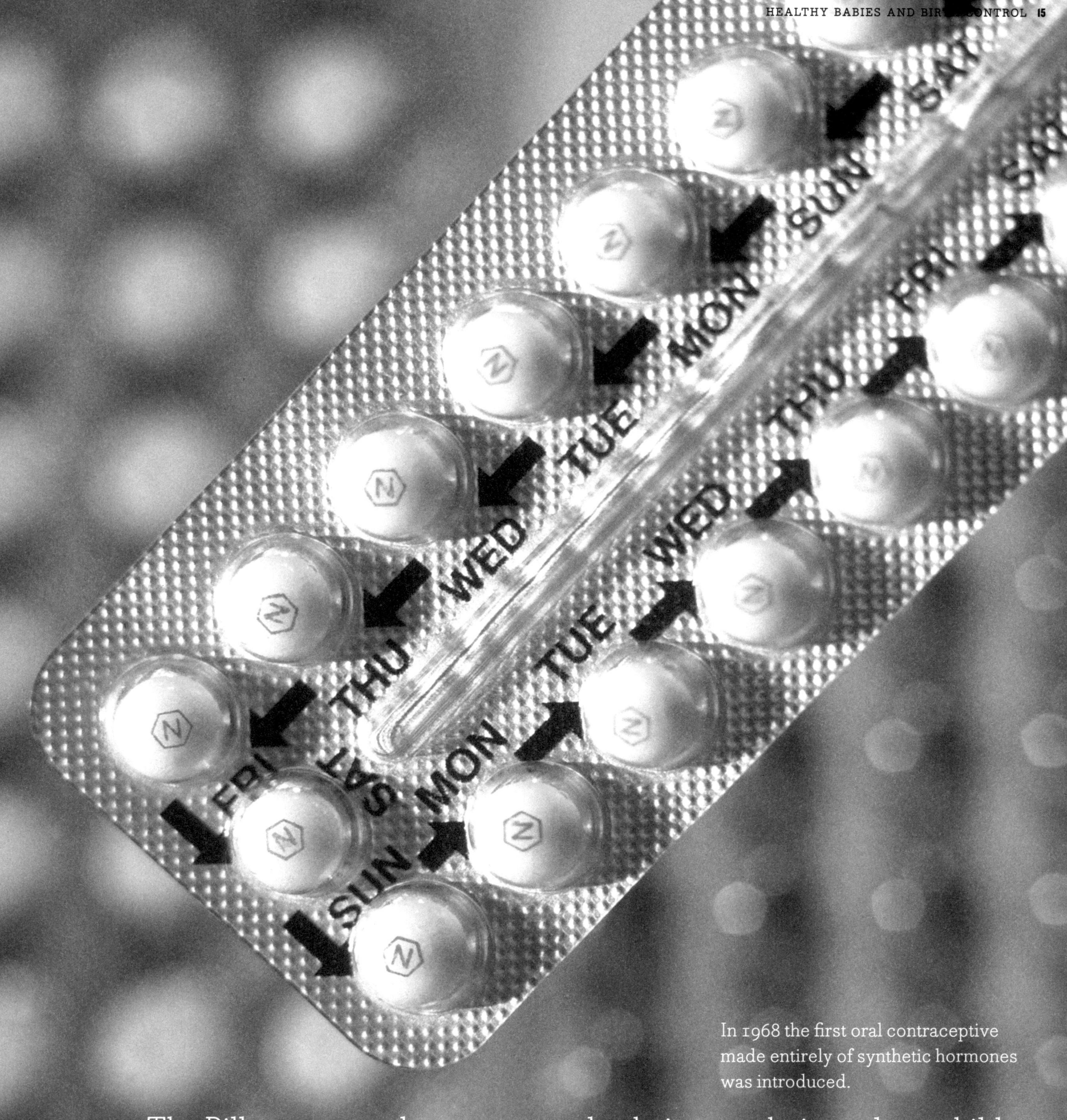

In 1968 the first oral contraceptive made entirely of synthetic hormones was introduced.

The Pill empowered women to make their own choices about children

Smith was one of the most financially astute academics of his generation, registering hundreds of patents worldwide. The Pill made him a multi-millionaire. And following his death he left $90 million for the University of Cambridge, the largest known gift by an individual to a UK university.

IMMACULATE CONCEPTION

On the night of 25 July 1978 medical history was created. A 5lb 12oz baby called Louise Joy Brown was born at 11:47pm by Caesarean section at Oldham General Hospital.

They were able to find a reliable way of fertilising a woman's egg in a laboratory test tube and then implanting the newly created embryo in a womb, where it developed into a baby nine months later.

The 1960s: We said, 'In the future, infertile couples will have babies'

Louise was the first 'test-tube baby', created by in-vitro fertilisation (IVF) where a mother's egg and father's sperm are fused outside the body in a laboratory.

University of Cambridge embryologist, Robert Edwards and the Oldham based gynaecologist, Patrick Steptoe, were the first to develop the IVF technique, to enable infertile women to have babies.

A quarter of a century later, it is estimated that worldwide one million people owe their lives to IVF. Doctors now routinely use the technique to help women who have blocked fallopian tubes or other complications preventing normal births. IVF babies, including Louise's sister Natalie, have gone on to conceive their own healthy babies naturally.

Modern-day infertility treatment now allows parents to have babies that are free from many of the debilitating inherited 'genetic' diseases that have plagued families for generations.

TREATING TINY PATIENTS

Making IVF treatment more affordable and accessible

Medical scientists, led by Robert Winston at Imperial College London, have developed a number of tests that enable doctors to select newly created embryos that do not contain the genetic abnormalities that can trigger life threatening conditions such as muscular dystrophy and haemophilia (where blood fails to clot).

These diseases can be avoided as they are associated with a particular sex. Scientists can now discover the sex of an embryo before it is implanted in the uterus. This means, for example, that parents can avoid having male children carrying fatal genetic disorders.

Winston and his team at Hammersmith Hospital have been responsible for many other world firsts that have improved the effectiveness and reliability of fertility medicine. They were the first group successfully to freeze ovarian tissue. And they created the first National Health Service-run IVF programme.

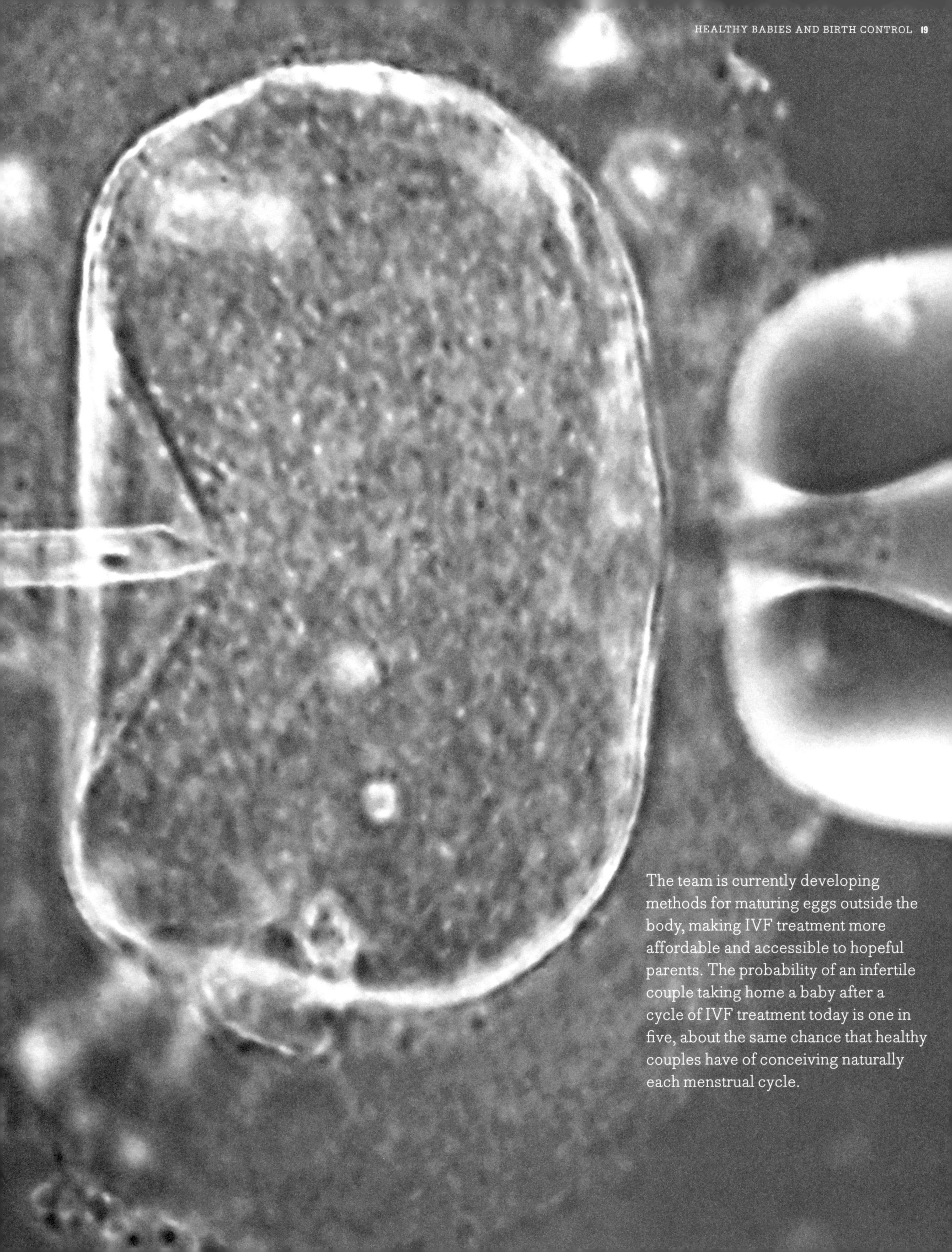

The team is currently developing methods for maturing eggs outside the body, making IVF treatment more affordable and accessible to hopeful parents. The probability of an infertile couple taking home a baby after a cycle of IVF treatment today is one in five, about the same chance that healthy couples have of conceiving naturally each menstrual cycle.

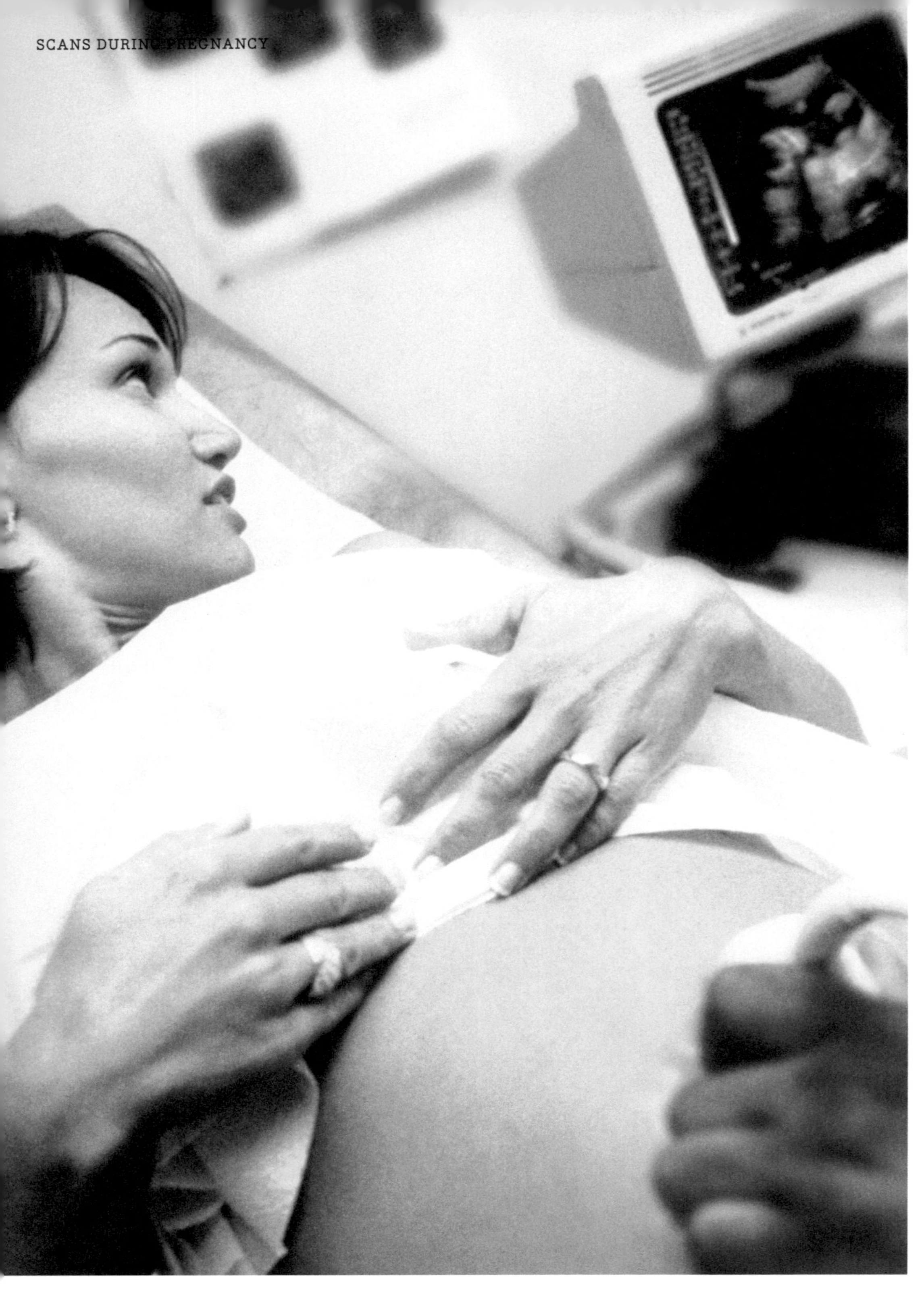

Ripples of 'ultrasonic' waves are transformed into a moving picture of a baby on a computer screen. The 'ultrasound scan' is now common practice in UK hospitals, allowing doctors to detect any early problems during pregnancy.

Ian Donald invented the use of ultrasound for unborn babies at the University of Glasgow 40 years ago.

Being able to see images of the foetus during the early stages of pregnancy has helped doctors to identify and treat problems that might otherwise put babies' lives at risk.

They can monitor the development of the baby, check whether there are any abnormalities, and determine its position inside the womb.

Scans have also improved our understanding of foetal development and the beginnings of human life.

Images of the foetus can help doctors to identify and treat problems

BABY PICTURES

Today many parents first see their child not in the hospital delivery room, but through images created from sounds, undetectable by the human ear, bounced off a baby's body in the womb.

SAFE SLEEPING

Two decades ago, 2,000 babies every year were dying suddenly and inexplicably while sleeping in their cots in the UK. Researchers are still trying to understand why cot deaths occur. But in the 1980s Peter Fleming and Jem Berry at the University of Bristol made a major breakthrough.

Research has prevented at least 100,000 infant deaths worldwide

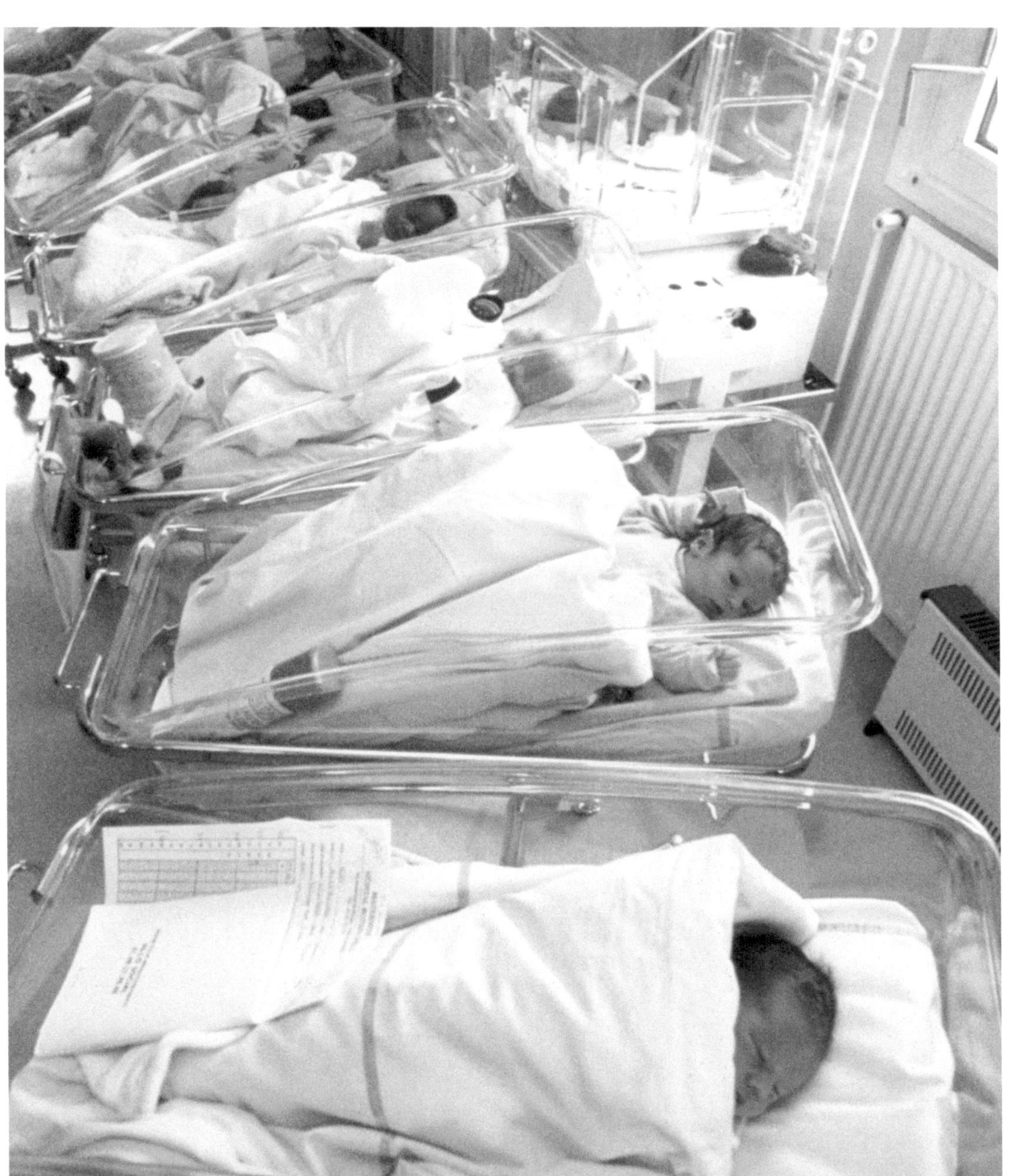

The researchers uncovered a link between the sleeping position of babies and the unexplained deaths. Their study showed that babies were more likely to suffer from 'sudden infant death syndrome' (SIDS) if they were sleeping face down, exposed to parental tobacco smoke, or were covered by too many blankets (particularly in warm rooms).

After a successful trial in Avon, the national 'Back to Sleep' campaign was launched, advising parents that babies should sleep on their backs. Within a month, the number of cot deaths had fallen by between 30 to 40%; after two years, by 60%. The research has prevented at least 10,000 infant deaths in the UK, and 100,000 worldwide.

EARLY DETECTION

In 1974, Nicholas Wald, then at the University of Oxford, discovered a way of predicting whether unborn babies are likely to have debilitating paralysing conditions, such as spina bifida and anencephaly (where the brain is small, or missing altogether).

Using a simple blood test from the expectant mother, the academics found that a protein produced in the growing baby's liver called alpha-fetoprotein (AFP) was higher in the bloodstream of women carrying babies with these disorders.

Around a thousand pregnancies in the UK every year are affected by such birth defects. They are caused by damaged nervous tissue in the spine of a growing foetus. One or more bones which form the spine fail to develop, leaving a gap for nervous tissue to become entangled. Spina bifida can lead to weakening or paralysis of the legs and poor control of urination and defecation. Anencephaly is always fatal.

In 1988 at St Bartholomew's Hospital Medical College (now part of Queen Mary, University of London) Wald and his colleagues developed another blood test using the same principles. He and fellow researchers showed that the measurement of three substances (including AFP) in a pregnant woman's blood could be used to screen women for Down's Syndrome pregnancies as well. The test is now known as the Barts or triple test.

Three years later, Wald was able to move from screening and diagnosis to actual prevention. He showed that by taking a simple vitamin supplement, folic acid, expectant mothers could reduce the risk of having a child with anencephaly or spina bifida by about 75 per cent.

Antenatal screening for anencephaly and spina bifida is now routine throughout the world as is the triple test. Doctors and health authorities meanwhile advise women to take folic acid supplements before and during pregnancy.

About 1,000 UK pregnancies annually are affected by defects

SMALLER AND WEAKER

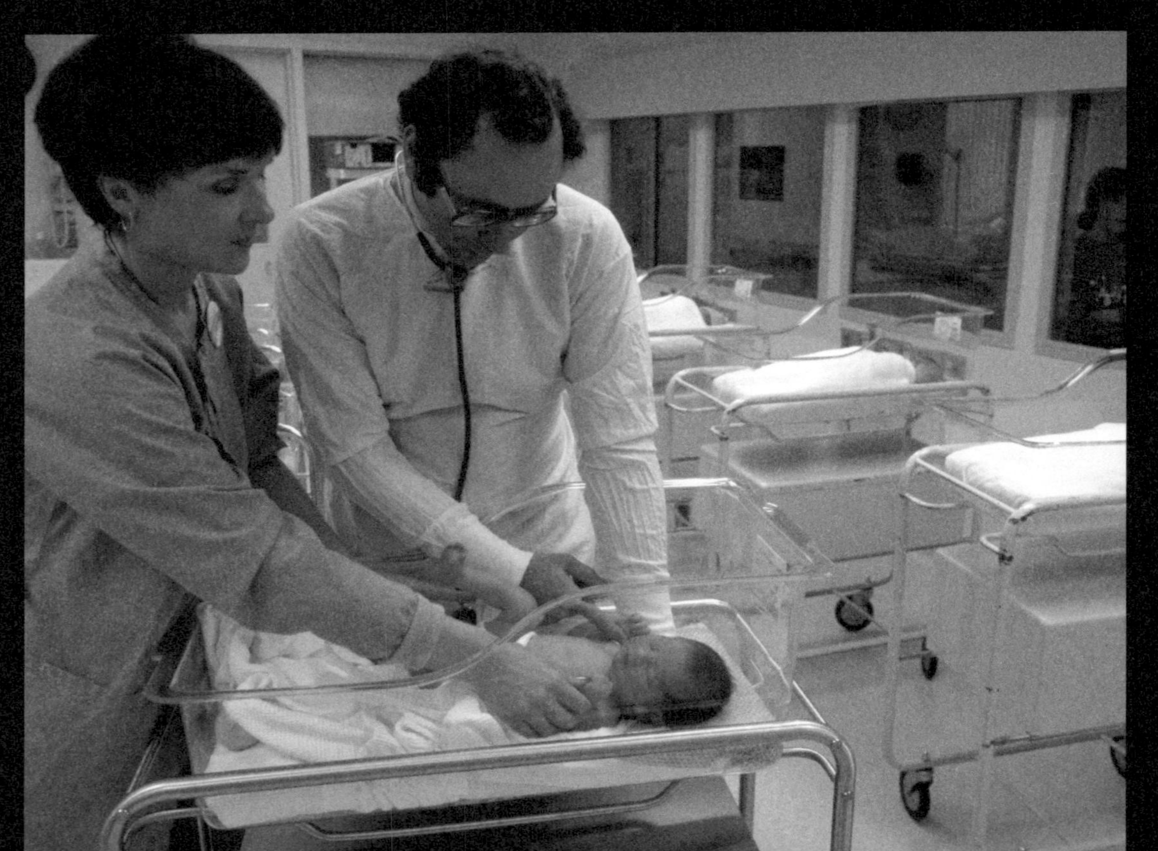

Smoking damages not only your own health, but also that of your unborn child. The advice seems obvious now. Yet the evidence only emerged after a ground-breaking study by UK academics a little over 30 years ago.

During the 1970s Harvey Goldstein and Neville Butler, then based at the National Children's Bureau in London, studied the details of 17,000 babies born in 1958. Studying children from the National Child Development Study, they discovered that babies with mothers who smoked were lower in weight by an average of 200g than other babies. Smoking cuts down the oxygen reaching the baby when it is the womb making it smaller and weaker.

Official warnings on cigarette packets and tobacco adverts have publicised the message ever since, helping dramatically to reduce dramatically the number of women who smoke when pregnant. Fewer babies are now born with small organs and the developmental problems caused by smoking.

The National Child Development Study continues today at the Institute of Education, part of the University of London.

Babies with mothers who smoked were lower in weight

BOB EDWARDS AND PATRICK STEPTOE

THE PARTNERS WHO PRODUCED THE FIRST TEST-TUBE BABY

Edwards was warned that Steptoe was a dangerous maverick

Bob Edwards has called it his eureka moment, when he first came across the name of Patrick Steptoe in the medical journal, *The Lancet* in 1968. He was warned that Steptoe was a dangerous maverick. But when they met a partnership began that led a decade later to one of the defining moments of modern medicine, the first 'test-tube baby'.

Edwards, an embryologist based at the University of Cambridge, had achieved the first fertilisation of a human egg outside the body in 1967. But Edwards needed someone to help him retrieve eggs from ovaries so that he could continue his work.

Steptoe was the perfect partner: he had pioneered the laparoscopic technique, allowing female reproductive organs to be viewed directly by inserting a thin telescope through a tiny incision in the navel (now referred to as 'keyhole surgery').

The only problem was that Steptoe was based in Oldham. Edwards spent much of the next several years making the 200-mile journey from Cambridge to the north, as the pair perfected the techniques of collecting eggs from women, fertilising them in the laboratory and then replacing them in the women.

The obstacles to the research were not just scientific. Edwards contacted the Medical Research Council to back a proposed move south for Steptoe to take a consultant's position near Cambridge. But the Council refused, saying that both laparoscopy and the implantation of eggs fertilised outside the womb were too hazardous. Edwards and Steptoe later said they 'felt sick reading that letter'.

But by the mid-1970s the pair were ready to attempt pregnancies using their techniques. And on the night of 25 July 1978, the world was introduced to the 5lb 12oz baby called Louise Joy Brown, the first baby to born by in-vitro fertilisation (IVF).

The birth was heralded as one of the most important medical advances of the 20th century. The Vatican, however, warned that it was 'an event that can have very grave consequences for humanity'.

Steptoe died in 1988, just after the 1,000th test tube baby was born, and just before he was to be knighted. Edwards, Emeritus Professor of Human Reproduction at the University of Cambridge, now runs the journal, *Reproductive Bio Medicine Online*. Louise Brown meanwhile, now a postal worker, celebrated her 25th birthday last year, one of more than a million babies born worldwide through IVF.

HEALTHIER AND LONGER LIVES

If you go to hospital you will more than likely benefit from UK academic research. Academics invented the impressive array of powerful probes that doctors now use to detect problems and diseases.

Ultrasound scans can highlight fragile bones in the old; magnetic resonance images can show the beating of the heart and even brain activity without touching the body at all; light emitting chemicals can provide super-quick tests for allergies, cancer and HIV; and keyhole surgery allows operations to be performed without having to cut the body open.

New technological innovations developed by researchers meanwhile have extended the lives of thousands. Complete hip replacements lasting more than 15 years, electronic pacemakers that vary the rate of your heartbeat, and portable defibrillators to resuscitate you if you suffer a heart attack, are all, these days, standard kit for the medical profession. Future wonders are set to include the first needle-free injection.

Major public health campaigns have also been inspired by the findings of medical research. It was academics who first dared to suggest the link between lung cancer and smoking, and who first discovered the benefits of fluoride for children's teeth.

Studies based in the UK can impact on the world: one Edinburgh study has led to a global vaccine for Hepatitis B, one of the most prolific killers across the globe. A novel artificial cow developed by researchers meanwhile has all but eradicated one of Africa's most harmful pests, the tsetse fly.

A third of women and one in 12 men over the age of 50 develop a condition called osteoporosis in which bones weaken and can break easily.

UNDER INVESTIGATION

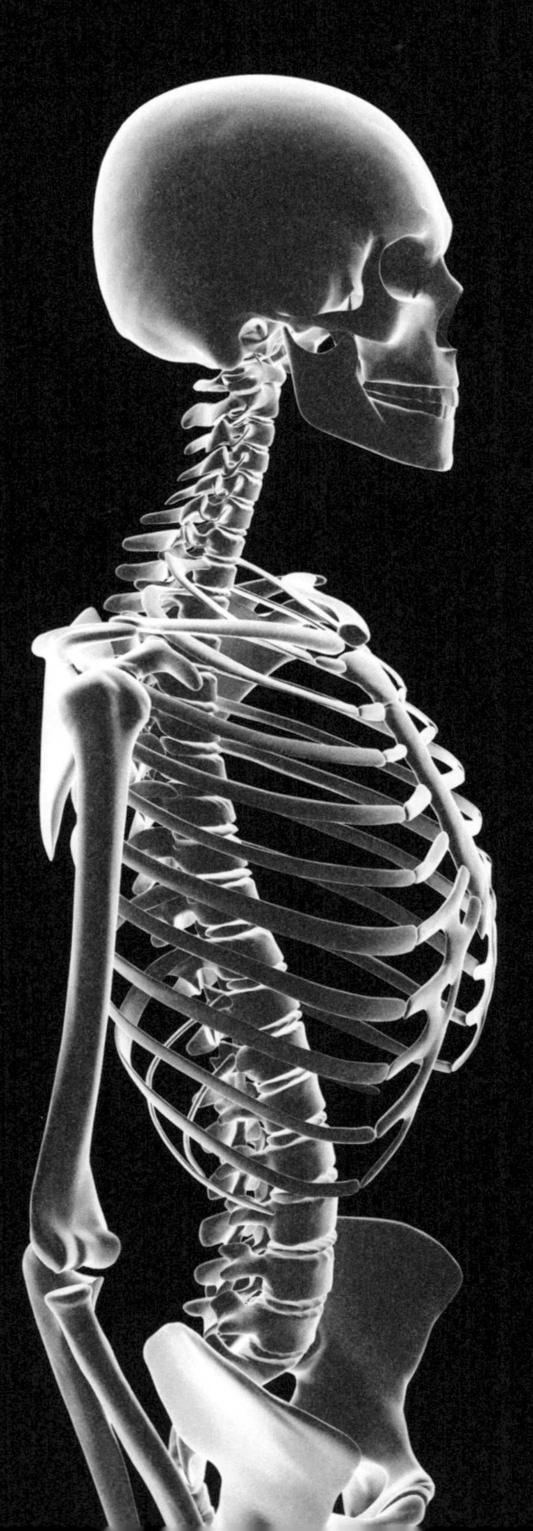

Sufferers get painful and disabling fractures, particularly in the wrist, hip and spine.

In the 1980s Chris Langton at the University of Hull was the first to develop an early detection system for the condition utilising 'ultrasonic' waves. These sounds are inaudible to the human ear, but can be reflected off bodies to reveal the structure and strength of bones in people likely to suffer from osteoporosis.

Using a bath of water, the ultrasound test identifies bone density by measuring how much of the ultrasound beam is absorbed by the bone under investigation.

Langton later helped the Royal Veterinary College to develop a method for identifying bone fractures in thoroughbred horses.

Chris Langton developed an early detection system for osteoporosis

The work had an eventual spin-off for humans as well; Langton subsequently developed the first successful 'dry' system for people in 1992.

About 12,000 detection machines have now been sold worldwide.

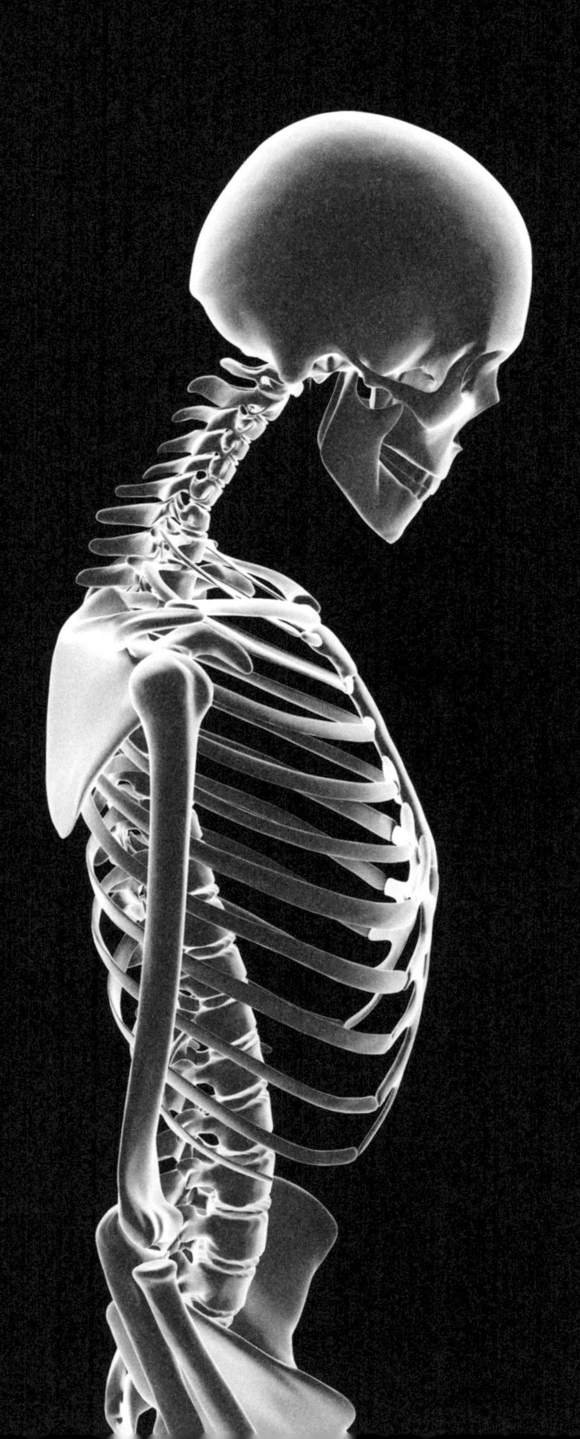

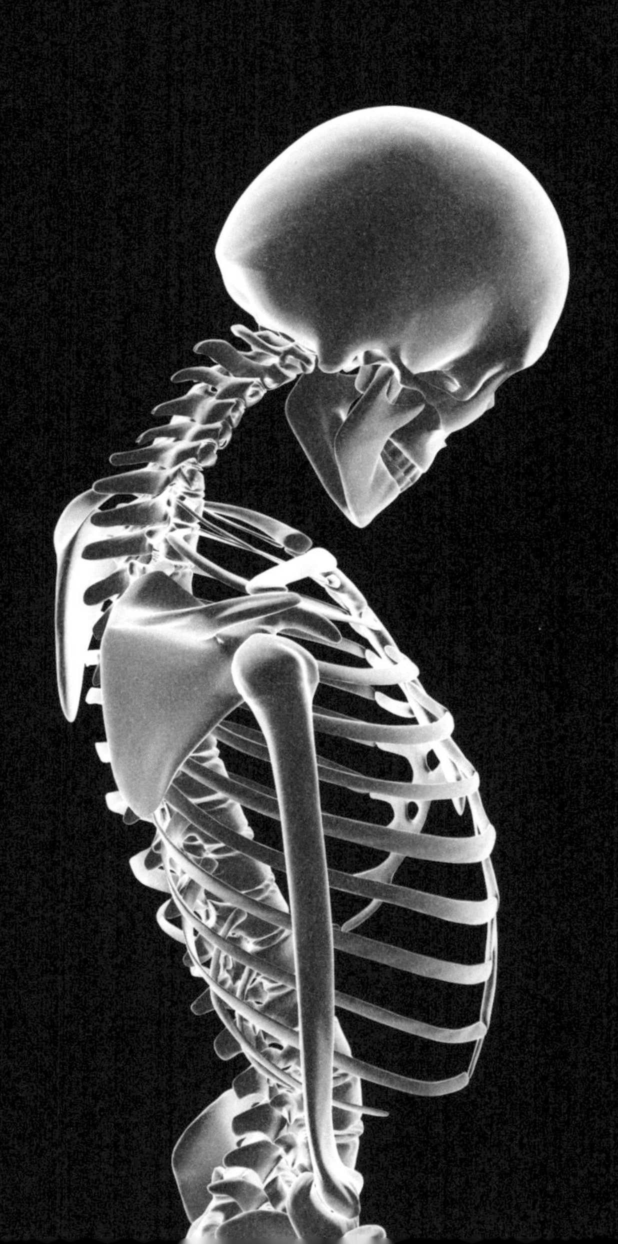

Magnetic Resonance Imaging allows doctors to monitor the beating of the heart, the flow of blood through the body and even brain activity without touching the body at all.

SUPER VISION

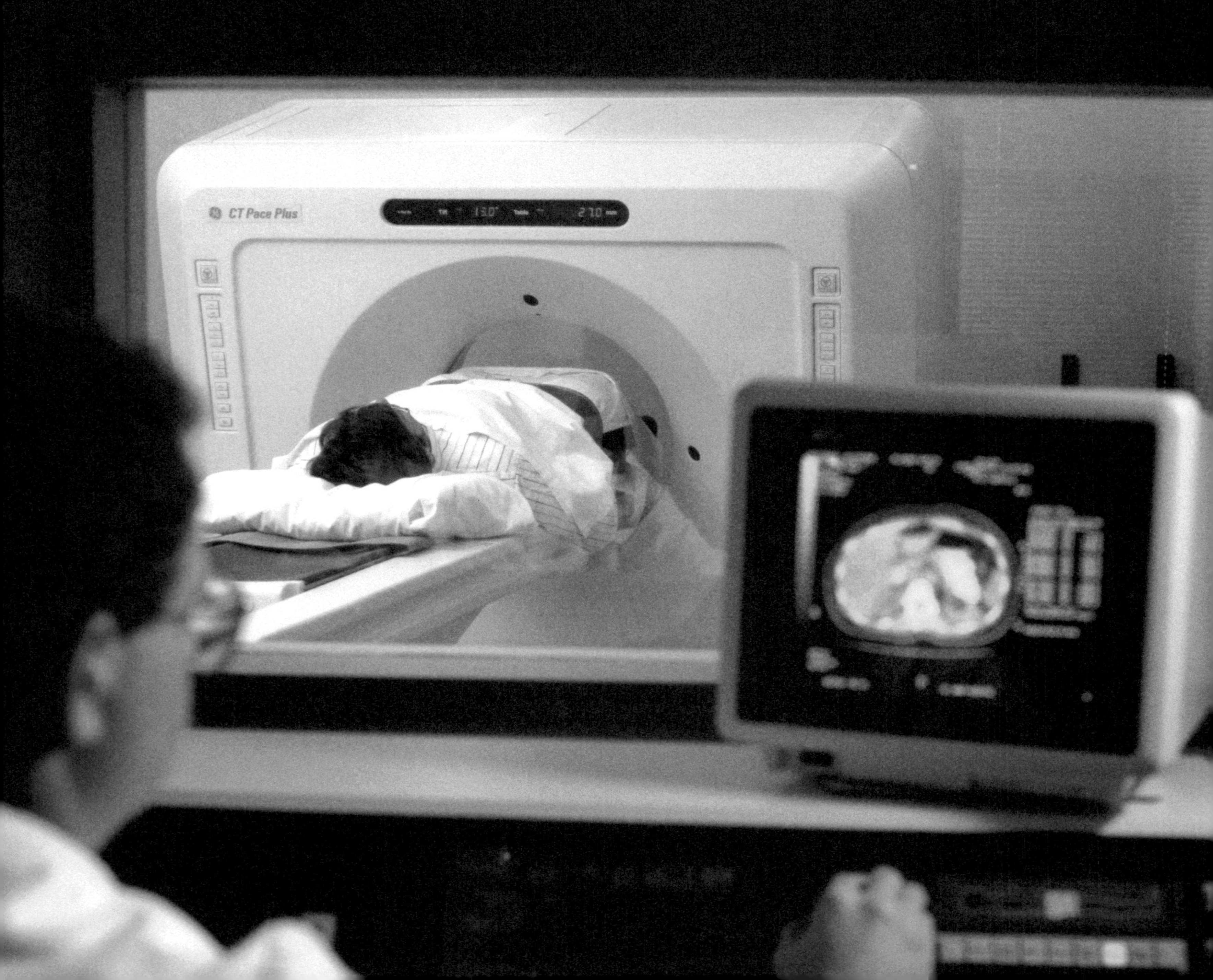

MRI scanners replace the need for riskier surgical procedures

The technique was developed by a number of UK academics during the 1970s and 1980s. In 1976 Peter Mansfield at the University of Nottingham was the first to publish a successful MRI scan of a living human body part – a finger.

During the 1980s John Mallard at the University of Aberdeen took MRI another step forward when he discovered a technique, known as spin-warp imaging, that could produce three-dimensional images unaffected by the movement of patients.

MRI scans produce a map of the water content of the parts of the body using magnetic and radio waves. It is based on a phenomenon discovered by scientists in the 1930s, called nuclear magnetic resonance, in which magnetic fields and radio waves cause atoms to give off tiny radio signals. The MRI scanners convert the signals into visual images.

Approximately 15,000 MRI scanners are now used in hospitals around the world, replacing the need for riskier surgical and X-ray procedures. The machines are standard kit for doctors detecting neurological diseases such as stroke, cancer, multiple sclerosis and Alzheimer's disease.

Sir Peter was awarded the 2003 Nobel Prize for Medicine (joint with Paul Lauterbur) for his work in developing the concept of MRI and for pioneering ultra-high speed imaging techniques.

ILLUMINATING REACTIONS

Scientists have exploited a phenomenon called 'chemiluminescence' – where chemicals emit light during reactions – to develop faster and more accurate tests for allergies, anaemia, cancer and HIV.

Chemiluminescent chemicals can be attached to proteins such as antibodies or to sequences of genetic material. These 'labelled' reagents can then be used to probe for molecules connected with diseases.

The technology, which was first developed by a group of scientists at the former University of Wales College of Medicine (now Cardiff University School of Medicine) in the early 1980s, is now in use in laboratories all around the world.

It is now possible to identify the presence of a few virus particles in a blood sample. As a consequence, the presence of HIV can be identified within a few days of a person becoming infected, whereas conventional methods are unable to detect the disease for three to four weeks.

It is now possible to identify a few virus particles in a blood sample

Methods are so powerful that they can even be used to detect the tiny sequence variations (polymorphisms) or mutations in DNA that induce inherited disorders or make certain individuals more susceptible to a particular disease.

It is estimated that more than 50 million tests are performed annually on a worldwide basis.

When Harold Hopkins and Narinder Singh Kapany showed in 1954 how a bundle of tiny pin-like glass fibres allowed light and images to be transmitted along them even when they were curved, few could have imagined the impact the device would have for surgical operations.

SEEING INSIDE

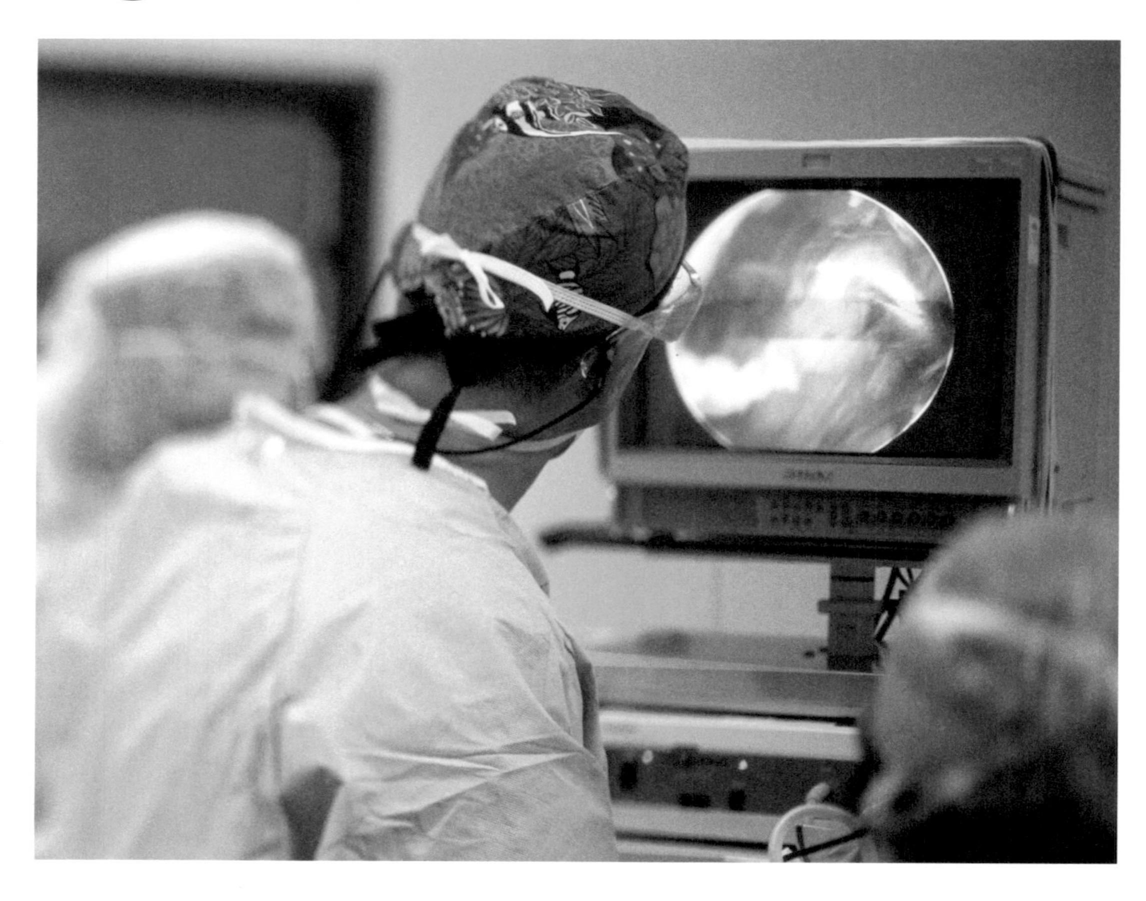

Fibre optics enabled doctors to operate without cutting the body open

The academics had invented fibre optics. The discovery revolutionised surgery, enabling doctors to see inside the human body, and operate without having to cut the body open to access internal organs.

Based at Imperial College London, Hopkins, who later moved to the University of Reading, designed the first flexible endoscope. The long, thin tube is inserted into the body, allowing doctors to peer through our passageways to examine and inspect organs, joints or cavities.

Before its invention, most internal conditions could only be diagnosed or treated through open surgery.

Thanks to the research, doctors now routinely perform 'keyhole' surgery, operating inside the body to treat diseases. The endoscope has transformed virtually every type of surgery. Many diseases, including some cancers, gallbladder stones and hernias, that were once common and difficult to treat, are now dealt with through minor surgery.

PACE OF CHANGE

During the 1950s electronic pacemakers were developed for people whose hearts beat too slowly.

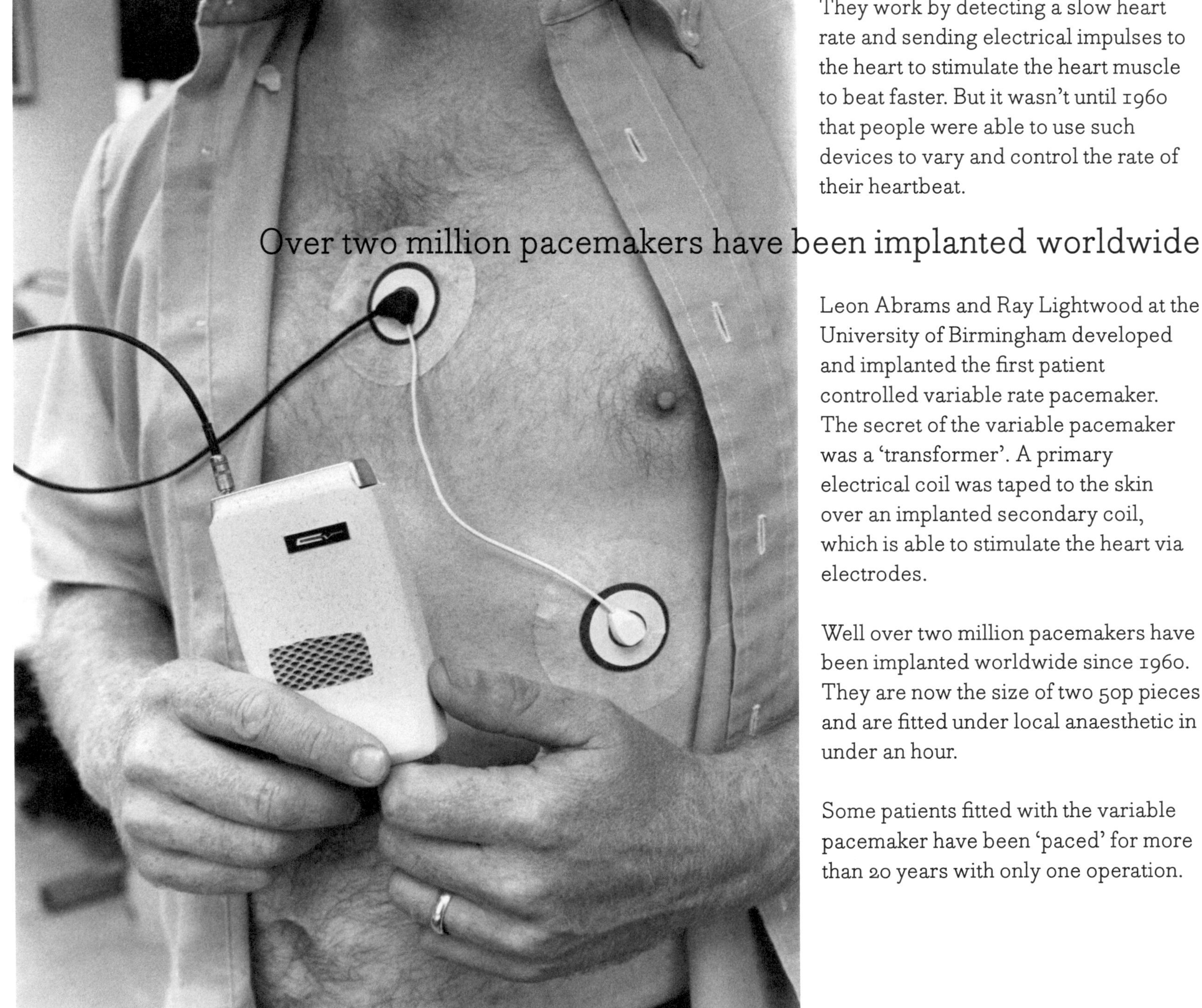

They work by detecting a slow heart rate and sending electrical impulses to the heart to stimulate the heart muscle to beat faster. But it wasn't until 1960 that people were able to use such devices to vary and control the rate of their heartbeat.

Over two million pacemakers have been implanted worldwide

Leon Abrams and Ray Lightwood at the University of Birmingham developed and implanted the first patient controlled variable rate pacemaker. The secret of the variable pacemaker was a 'transformer'. A primary electrical coil was taped to the skin over an implanted secondary coil, which is able to stimulate the heart via electrodes.

Well over two million pacemakers have been implanted worldwide since 1960. They are now the size of two 50p pieces and are fitted under local anaesthetic in under an hour.

Some patients fitted with the variable pacemaker have been 'paced' for more than 20 years with only one operation.

Since then Neil Jenkins, Andrew Rugg-Gunn and John Murray, based at the University of Newcastle, have compared the dental health of children living in fluoride areas with children living in adjacent low fluoride areas. They found that higher levels of fluoride in water were linked to fewer incidents of tooth decay among children.

Today 90 per cent of toothpastes sold in the UK contain fluoride, and in many parts of the country it is added to drinking water. Tooth decay in children meanwhile has fallen dramatically.

As with many areas of research however, the academic debate over the use of fluoride continues. Often there are risks as well as benefits associated with medical advances.

Today 90 per cent of toothpastes sold in the UK contain fluoride

A school dentist remarked that the children from South Shields (a natural fluoride area) had much better teeth than those of local children.

Recent studies have suggested that too much fluoride in water together with fluoride toothpaste and fluoride supplements can have damaging side-effects, such as unsightly marking on teeth (dental fluorosis).

It was the evacuation of children from the South Shields area to the Lake District during the Second World War that first hinted at the benefits of fluoride for teeth.

WAR ON TOOTH DECAY

In 1961, John Charnley achieved the Holy Grail of hip specialists across the world – the first complete hip replacement.

HIP DESIGN

The University of Manchester professor performed the first operations to replace whole hips at Wrightington Hospital in Wigan. Before Charnley's pioneering work, surgeons were only able to replace the different parts of the hip joint.

Complete hip replacement allows those with failing hip joints to walk more comfortably and without pain. The surgical procedure involves replacing the entire hip joint, composed of two parts, the hip socket, a cup shaped bone in the pelvis, and the 'ball' or head of the thigh bone (femur).

The operation is now standard practice in hospitals. In the UK, 40,000 people every year now walk again following complete hip operations.

40,000 people a year now benefit from hip operations

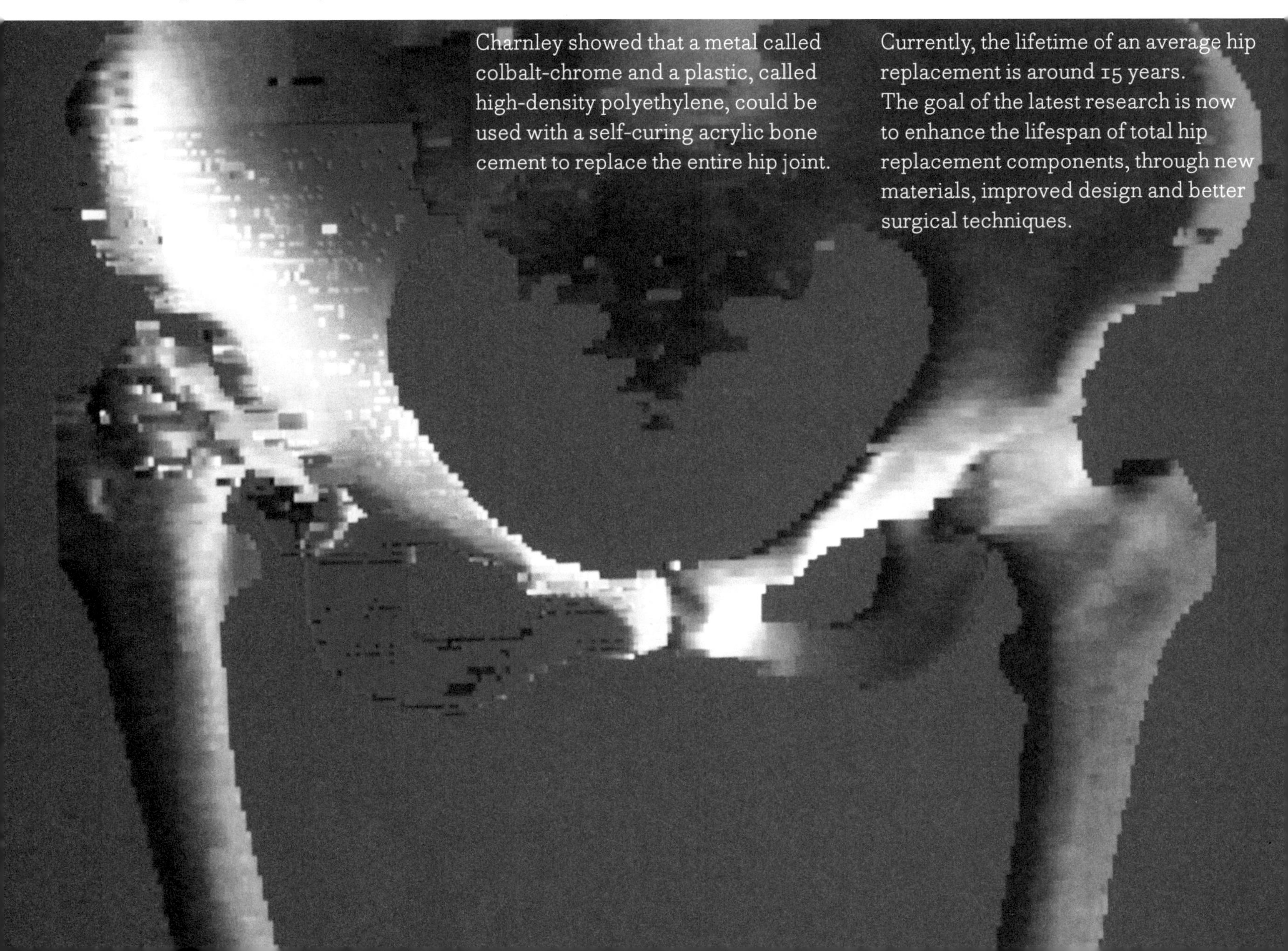

Charnley showed that a metal called colbalt-chrome and a plastic, called high-density polyethylene, could be used with a self-curing acrylic bone cement to replace the entire hip joint.

Currently, the lifetime of an average hip replacement is around 15 years. The goal of the latest research is now to enhance the lifespan of total hip replacement components, through new materials, improved design and better surgical techniques.

SHOCK TACTICS

There was time when heart attack victims stood little chance of survival outside hospitals. But the portable defibrillator, developed in the early 1970s by Frank Pantridge at Queen's University of Belfast, has since saved thousands of lives.

It allows the emergency services to resuscitate people suffering heart attacks as soon as possible, wherever they are.

Defibrillators work by a sending a jolt of electricity to restart the heart, but work only if applied within minutes of a heart attack. Pantridge developed the portable defibrillator, an electric device that can be carried around by emergency services and charged with a car battery.

Over the last 50 years, portable defibrillators have reduced the number of deaths from heart disease. In 1999 the Government announced a 'Defibrillators in Public Places' initiative to provide 700 defibrillators in busy public places, such as major railway stations, airports and shopping centres.

One quarter of the 57,000 deaths due to heart attacks in the country occur in public places.

One quarter of deaths due to heart attacks occur in public places

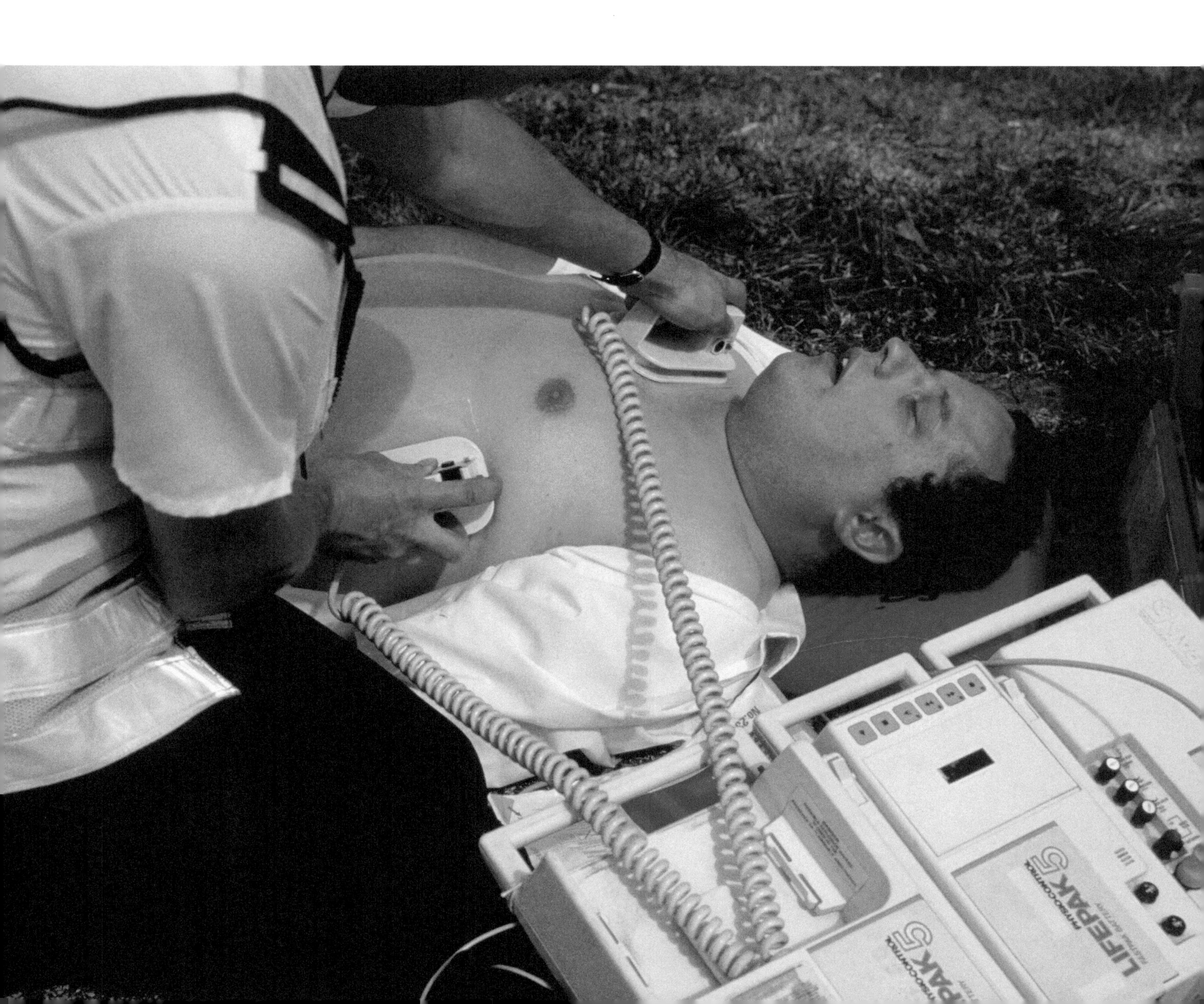

JABS ARE A BREEZE

Imagine receiving an injection where instead of feeling the prick of a needle, all you sense is a light breeze on the skin.

The needle-free powder injection system utilises helium gas

The needle-free injection is in fact in development, and it is only a matter of time before it may be used to improve the lives of children and adults who fear injections.

It will also help doctors and nurses who run the risk of being accidentally infected with diseases such as HIV or Hepatitis B from contaminated needles.

In 1993 Brian Bellhouse at the University of Oxford invented a way of giving vaccinations and other treatments without a needle. The needle-free powder injection system utilises helium gas to deliver tiny dry pharmaceutical particles into the skin.

The low density gas is reflected from the skin, but the microscopic particles penetrate the epidermis, triggering a response from the body's immune system. All patients feel is a light breeze on their skin as the vaccine is administered.

It has been shown in clinical trials that the hand-held system may be particularly effective in the delivery of vaccines. This is because the vaccine is delivered direct to the immune system through the epidermis.

Conventional vaccines for 'flu in dry powder form are already involved in clinical trials. A new needle-free 'flu vaccination could be available to the public in the next few years.

Four fifths of adu
And despite risi
experts suspecte
as the cause.

DEAD

Few people, however,
them, and they set out
the findings by another
They undertook one of
major 'epidemiology' s
sending questionnaires
British doctors asking
smoking habits.

Responses were gathe
34,000 male doctors w
contributed to follow-u
next 50 years. Respons
6,000 female doctors c
similarly for 20 years.

The results were concl
They confirmed that pe
got the disease unless
been smoking.

1973–2003: smoking-related deaths amon

Then, working for the Medical Research Council, two academics Austin Bradford Hill and Richard Doll, published a ground breaking study that found that 0.5 per cent of men with lung cancer were life-long non-smokers compared with five per cent of men of the same ages in the general population. This, they thought, together with other evidence proved that smoking was the principal cause of the disease.

The work moved to the
of Oxford when Doll w
Regius Professor of Me
and continued later in
Trial Service Unit with

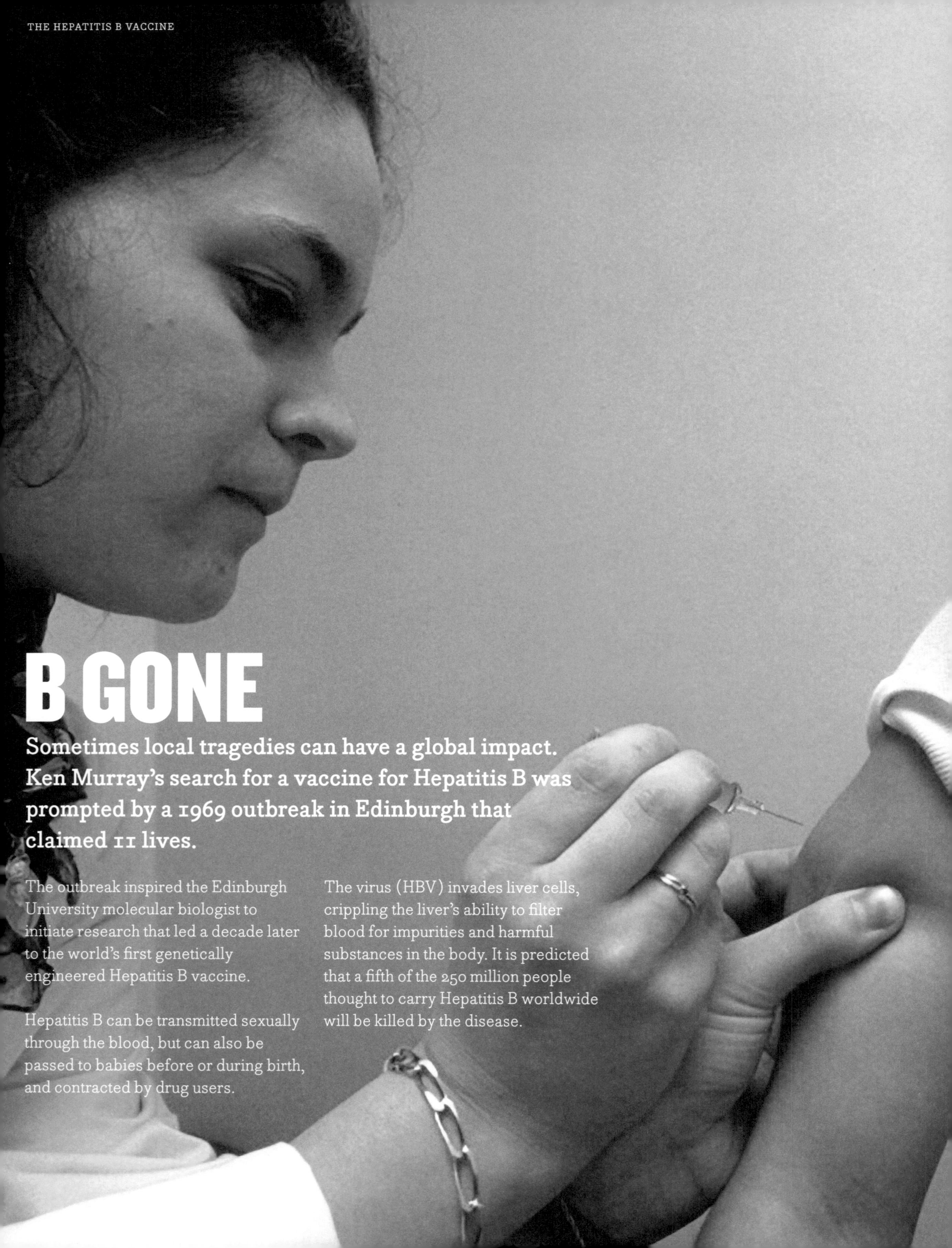

B GONE

Sometimes local tragedies can have a global impact. Ken Murray's search for a vaccine for Hepatitis B was prompted by a 1969 outbreak in Edinburgh that claimed 11 lives.

The outbreak inspired the Edinburgh University molecular biologist to initiate research that led a decade later to the world's first genetically engineered Hepatitis B vaccine.

Hepatitis B can be transmitted sexually through the blood, but can also be passed to babies before or during birth, and contracted by drug users.

The virus (HBV) invades liver cells, crippling the liver's ability to filter blood for impurities and harmful substances in the body. It is predicted that a fifth of the 250 million people thought to carry Hepatitis B worldwide will be killed by the disease.

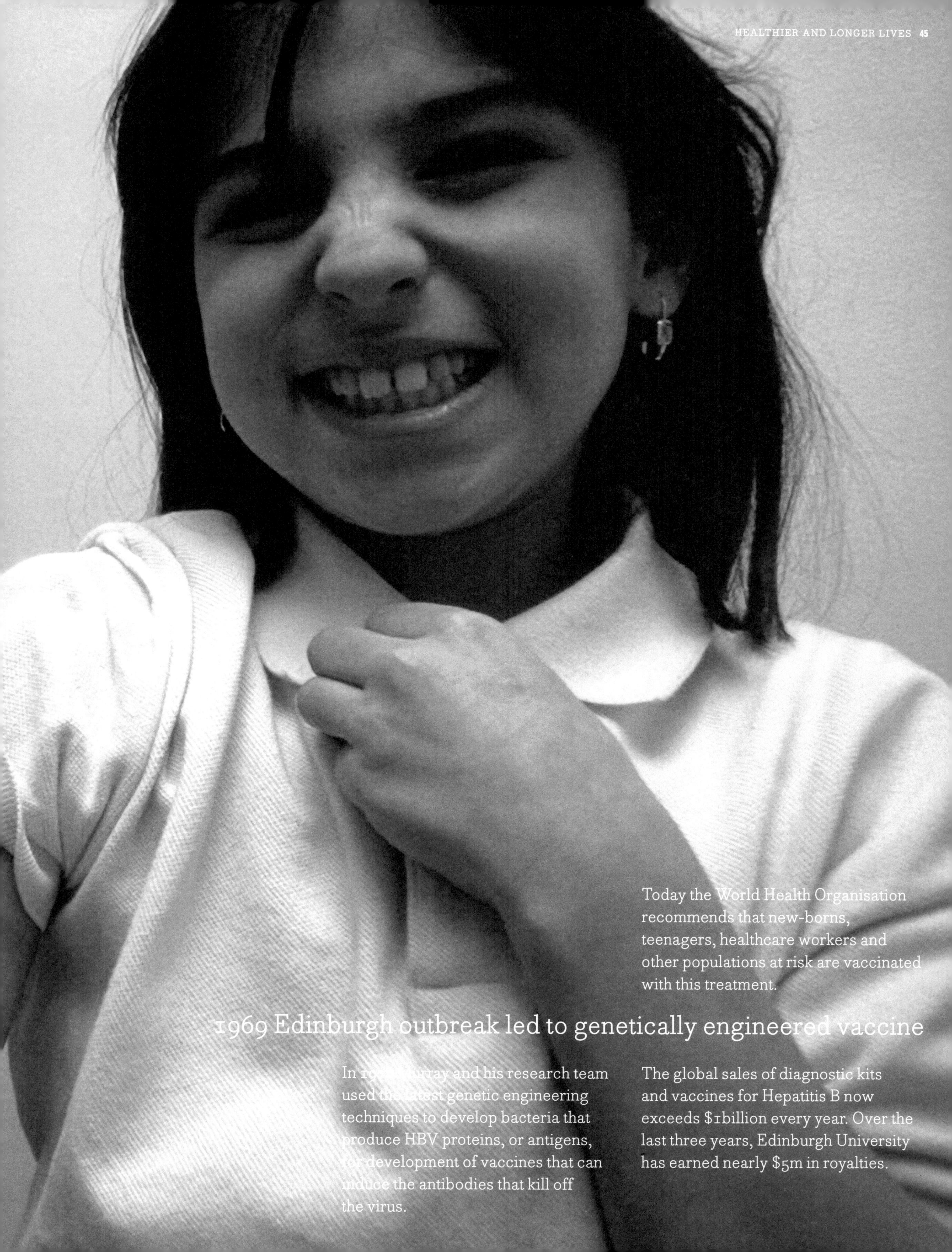

Today the World Health Organisation recommends that new-borns, teenagers, healthcare workers and other populations at risk are vaccinated with this treatment.

1969 Edinburgh outbreak led to genetically engineered vaccine

In [illegible] Murray and his research team used the latest genetic engineering techniques to develop bacteria that produce HBV proteins, or antigens, for development of vaccines that can induce the antibodies that kill off the virus.

The global sales of diagnostic kits and vaccines for Hepatitis B now exceeds $1billion every year. Over the last three years, Edinburgh University has earned nearly $5m in royalties.

BOVINE INTERVENTION

One of Africa's most harmful pests, the tsetse fly, has been all but eradicated from parts of the continent thanks to a novel artificial cow developed by an international group of researchers, including scientists from the University of Greenwich.

Cases of nagana in Zimbabwe have plummeted to practically zero and have remained at this low level for the last five years, largely due to the use of artificial cows, of which there are now about 60,000 in use. The fake cows also act as an effective barrier to stop tsetse re-invading areas cleared of flies.

Not only are artificial cows highly successful in controlling tsetse, but their use also results in a dramatic reduction in the amount of insecticide necessary to control the pest.

Today: practically zero cases of sleeping sickness

The artificial cows attract tsetse – which can infect humans and cattle with fatal sleeping sickness – by emitting chemicals (kairomones) to mimic the smell of real cattle. The fake cattle are impregnated with insecticides that kill the tsetse attracted to them.

The cows were introduced to Zimbabwe in the mid-1980s, when thousands of cattle were infected with nagana (the equivalent to human sleeping sickness), transmitted by tsetse.

With only four artificial cows needed per square kilometre to ensure effective pest control, the use of insecticide is far more targeted than conventional widespread aerial and ground spraying, resulting in a greatly reduced environmental impact.

RICHARD DOLL
THE MAN WHO PROVED THE LINK BETWEEN SMOKING AND LUNG CANCER

Richard Doll blames drink for going into medicine. He failed to get a scholarship from the University of Cambridge to study mathematics because of a particularly bad hangover. Instead, the young student (who smoked until he was 37 years old) trained to be a doctor at St Thomas' Hospital, London.

That career decision has helped to save the lives of thousands over the last 50 years. Doll was one of the medical researchers who later revealed in 1950 the link between smoking and lung cancer.

In 1947, the Medical Research Council set up an investigation into the dramatic increases in cases of lung cancer in the UK. Most experts thought that atmospheric pollution from coal smoke was to blame, while others suspected road tar, which was known to contain several carcinogens. Few people at the time imagined smoking was the cause, even though the vast majority of adults at the time smoked cigarettes.

After serving on a hospital ship in the Mediterranean during the World War Two, Doll returned to St Thomas', but soon became disillusioned with the hierarchical nature of the medical profession, and sought instead to become a researcher.

For the MRC study, Doll joined Austin Bradford Hill and helped to interview thousands of lung cancer patients about their lifestyle and habits. The conclusion was clear to them: there was an irrefutable link between smoking and lung cancer.

They published their results in 1950. But it took seven years for the Government to accept the findings. And 40 more for the tobacco industry to do so. In the UK, the message appears to be getting through: smoking-related fatalities among British men in 2003 were less than half their level 30 years previously.

Doll is now known as one of the first pioneers of epidemiology, the field that investigates the causes of a disease through statistics rather than biology. He has since helped to prove that the risks of blood clots posed by contraceptive pills are outweighed by their benefits in protecting against some cancers, and that an aspirin a day protects against heart disease.

Doll blames drink for going into medicine

MEDICINE UNDER THE MICROSCOPE

During the last 50 years, UK scientists have peered deeper and deeper into the microscopic world of the molecules and mechanisms that underpin living organisms. Their discoveries have already had huge medical spin-offs (and helped in the fight against crime). The promise of future health benefits is seemingly limitless: improved diagnosis of diseases, new individualised drug treatments, human tissue repairs, and even eventual cures.

Modern molecular biology is said to have been born in 1953 when Crick and Watson unveiled the helical structure of DNA, the stack of molecules that passes the code of life from one generation to the next. But of equal importance has been the revealing of the complex molecular structures that make up the proteins involved in the chemical processes of life.

The study of how certain proteins, called antibodies, attack infections and diseases meanwhile has uncovered the secrets of the body's immune system and led to modern immunisation techniques.

More recently genetic science has come of age. Researchers have decoded the three billion letters or separate molecules that make up the 30,000 genes of human DNA. Scientists have identified the genes that regulate how our cells divide and proliferate. They have created the first cloned adult animal, Dolly the sheep. And in the future scientists hope to manipulate genes to grow the different cells that make up the organs and tissues of the human body.

The most famous application of modern genetics, however, has come not in fighting illness, but crime. Genetic fingerprinting, which can identify the DNA of individuals, has revolutionised criminal investigations, allowing police to identify people using the smallest samples left at crime scenes.

A blue plaque now hangs in the Cambridge pub, the Eagle, where two young scientists, James Watson and the late Francis Crick, announced to the bewildered locals on 28 February 1953 that they had found the 'secret of life'.

GENE GENIE

Half a century on from the unveiling of the double helix structure of deoxyribonucleic acid or DNA, we are still just beginning to comprehend its full impact on the world.

DNA is a spiral staircase of molecules that exists in all our cells and contains the recipe for living things and the characteristics that are passed on from one generation to another.

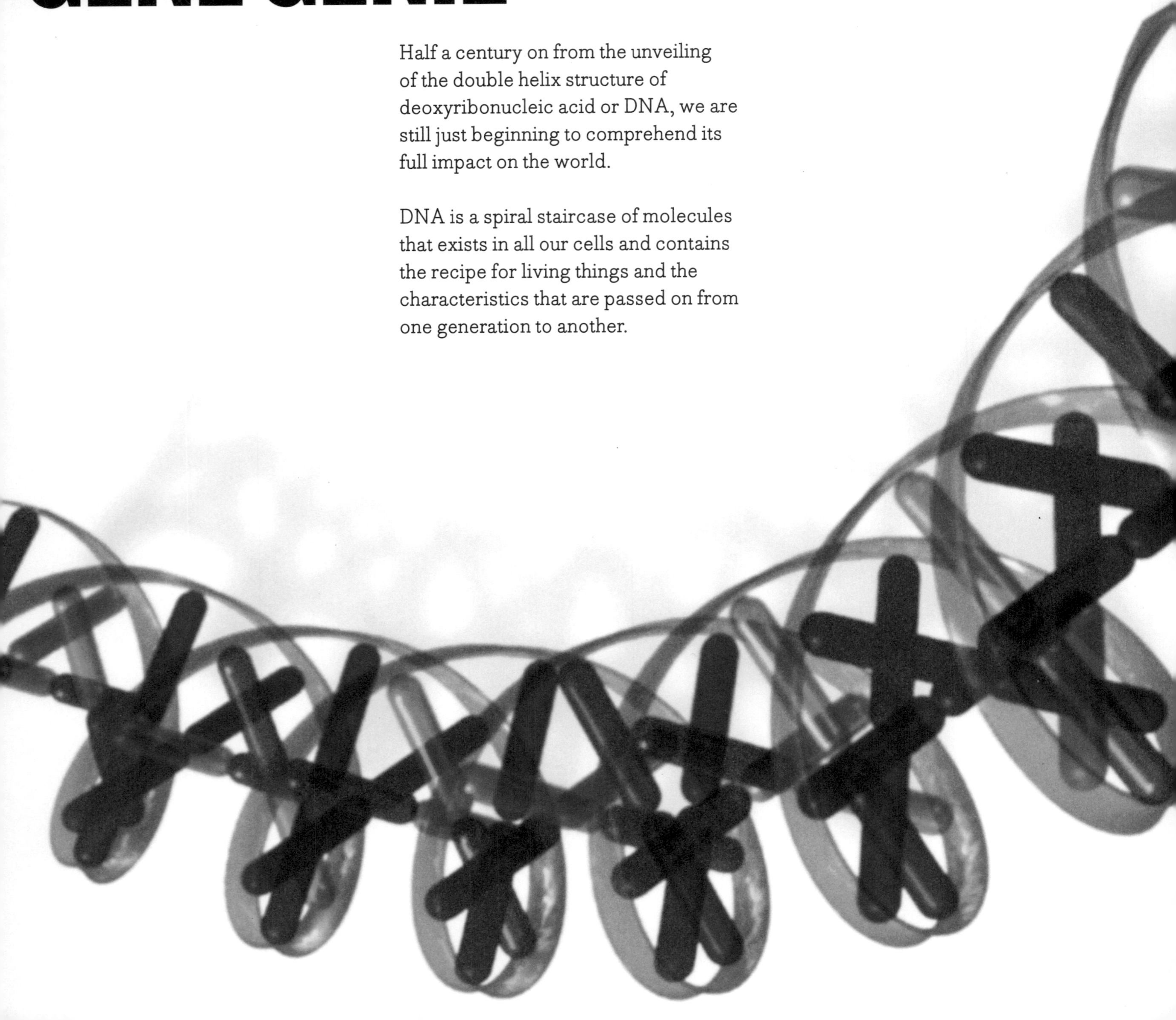

Watson and Crick's model – based on two strings of molecules called nucleotides connected to produce the helical ladder structure – enabled them to predict the basic functions of DNA and genes. Watson and Crick's discovery owed much to the X-ray images of DNA produced by Maurice Wilkins and Rosalind Franklin at King's College, London, which revealed the regular molecular building blocks that make up DNA.

Will genetic science produce personalised medicines and cures?

The discovery of the structure of DNA heralded the birth of modern molecular genetics. Since then scientists have been able to develop much more accurate ways of diagnosing conditions such as cystic fibrosis where the genetics of the disease is understood.

Researchers have created genetically engineered plants that are more resistant to attack from insects.

Scientists are also currently developing gene therapy, where new genes are introduced to compensate for so-called gene defects. Other spin-offs include DNA 'fingerprinting' techniques. The hope is that modern genetic science will one day produce personalised medicines and even cures for diseases.

Scientists working on the Human Genome Project 50 years on completed the decoding of the three billion letters or separate molecules that make up DNA. These combine to form about 30,000 genes that are the basis for human life. The challenge now is to understand how the genes interact together with their environment and via proteins to create the processes than underpin life.

In the next few decades, researchers aim to find out why some people are more susceptible to certain diseases and more receptive to some medicines but not others. It will help individuals be more aware of their own specific health risks. A longer-term aim is to create medicines that are tailored to the particular genetic make-up of an individual.

Working at the Cavendish Laboratory at the University of Cambridge during the 1950s, Perutz focused his research on haemoglobin, the protein responsible for making our blood red.

John Kendrew, one of Perutz's research students, unveiled the first structure of a protein, the muscle protein myoglobin, in 1957. Two years later, Perutz produced the structure for haemoglobin.

Thousands of protein structures have now been uncovered by researchers

Perutz was the first to show how the protein operates, taking up oxygen in the lungs and releasing it in the body's tissues. He found that mutations – small changes in the structure of the protein – could trigger diseases such as sickle-cell anaemia.

Perutz revealed the structure of the protein by using the latest X-ray experimental techniques. When placed in front of a beam of X-rays, proteins in the form of crystals produce a regular pattern of spots, the 'diffraction pattern'. Perutz was able to translate this pattern into a three-dimensional model of the protein.

Thousands of protein structures have now been uncovered by researchers, and there is much clearer understanding of their role in the body's functions, and their impact on disease.

MODEL EXAMPLE

Max Perutz pioneered the study of how proteins, the essential constituents of all living beings, work, illuminating for the first time their complex molecular structures.

The role played by insulin in the body has been known since the 1920s. Insulin is a protein secreted by the pancreas to allow cells to absorb sugar (and the energy it provides) from the blood.

VITAL PROTEINS

People suffering from early onset diabetes lack insulin, and have to receive injections of the protein.

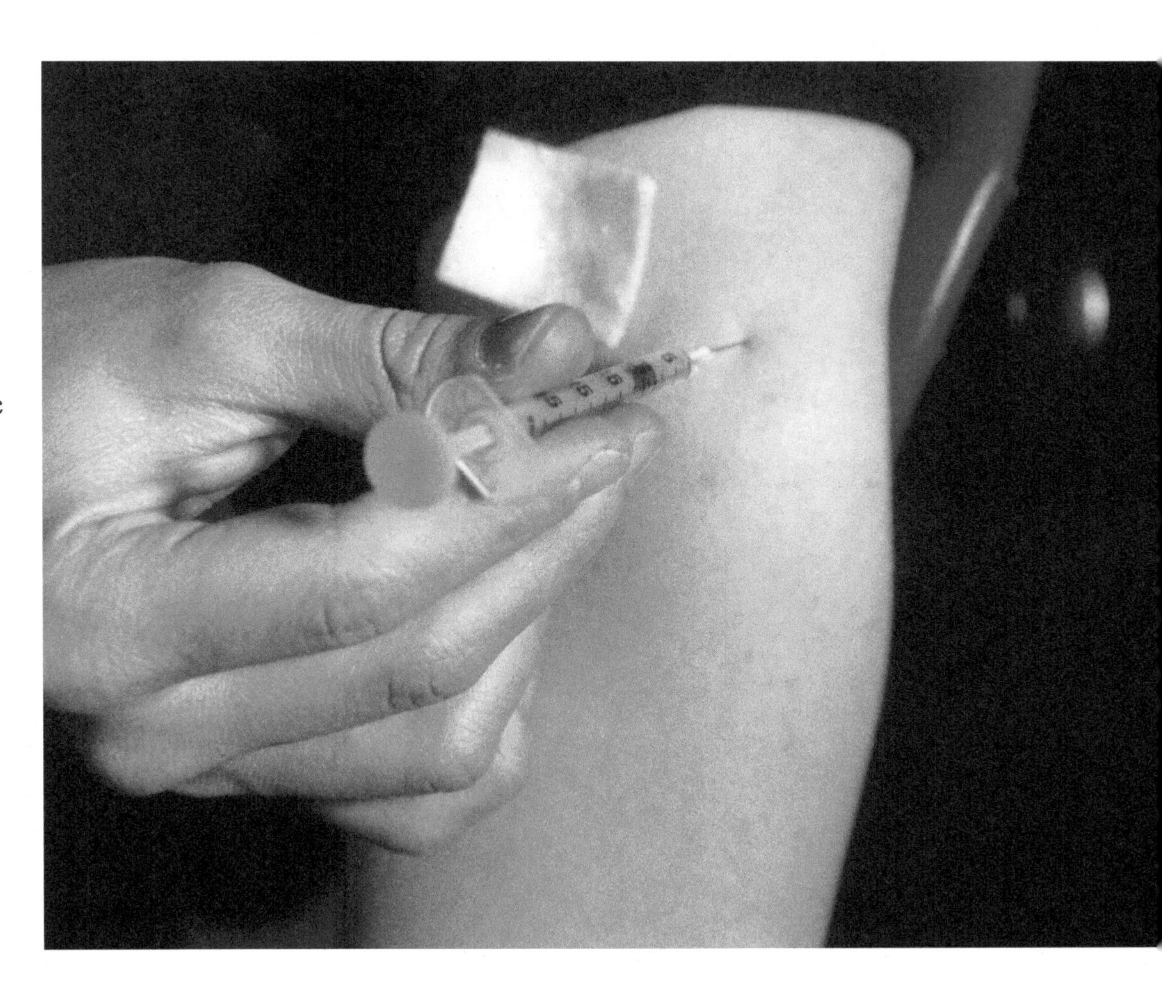

The actual molecular structure of insulin however was not uncovered until the 1950s. Working at the University of Cambridge, Fred Sanger revealed the exact order of the 51 basic building blocks, or amino acids, that make up the insulin molecule.

The discovery meant that insulin could be manufactured and made much more widely available.

Over 15 years, Sanger developed methods to identify individual amino acids using dye reagents and paper chromatography, and breaking down the molecule into fragments through oxidation.

Mutations could trigger diseases such as sickle-cell anaemia

Sanger found that the insulin molecule contains two different molecular chains, one with 31 amino acids, the other with 20. From the fragments Sanger was then able to reconstruct the molecules.

The techniques have since been applied by researchers to proteins in general, to determine their structure, and their role as key substances in the chemical processes of life.

They are called endorphins as they are the natural equivalent to drugs such as morphine and heroin. These drugs are addictive as they connect to the receptors in the brain intended for endorphins.

In 1975 Hans Kosterlitz and John Hughes at Aberdeen University were the first to show that endorphins are produced naturally by the body.

Endorphins turn off and on the internal systems that regulate pain and pleasure, especially during times of stress or injury. When the brain detects stress, it sends out endorphins, which bind at the receptor sites, alleviating pain.

LOVE AND PAIN

Endorphins are the reasons why people yearn for sex, laughter, touch and exercise. All these things trigger 'highs' within the body by activating naturally occurring chemicals in the brain.

At times of severe injury and chronic illness they ensure that our bodies' tissues are relaxed so that all the necessary antibodies can travel into the affected region in order to heal and repair. If the body is not in pain, a feeling of euphoria occurs.

Endorphins are the natural equivalent to drugs such as morphine and heroin

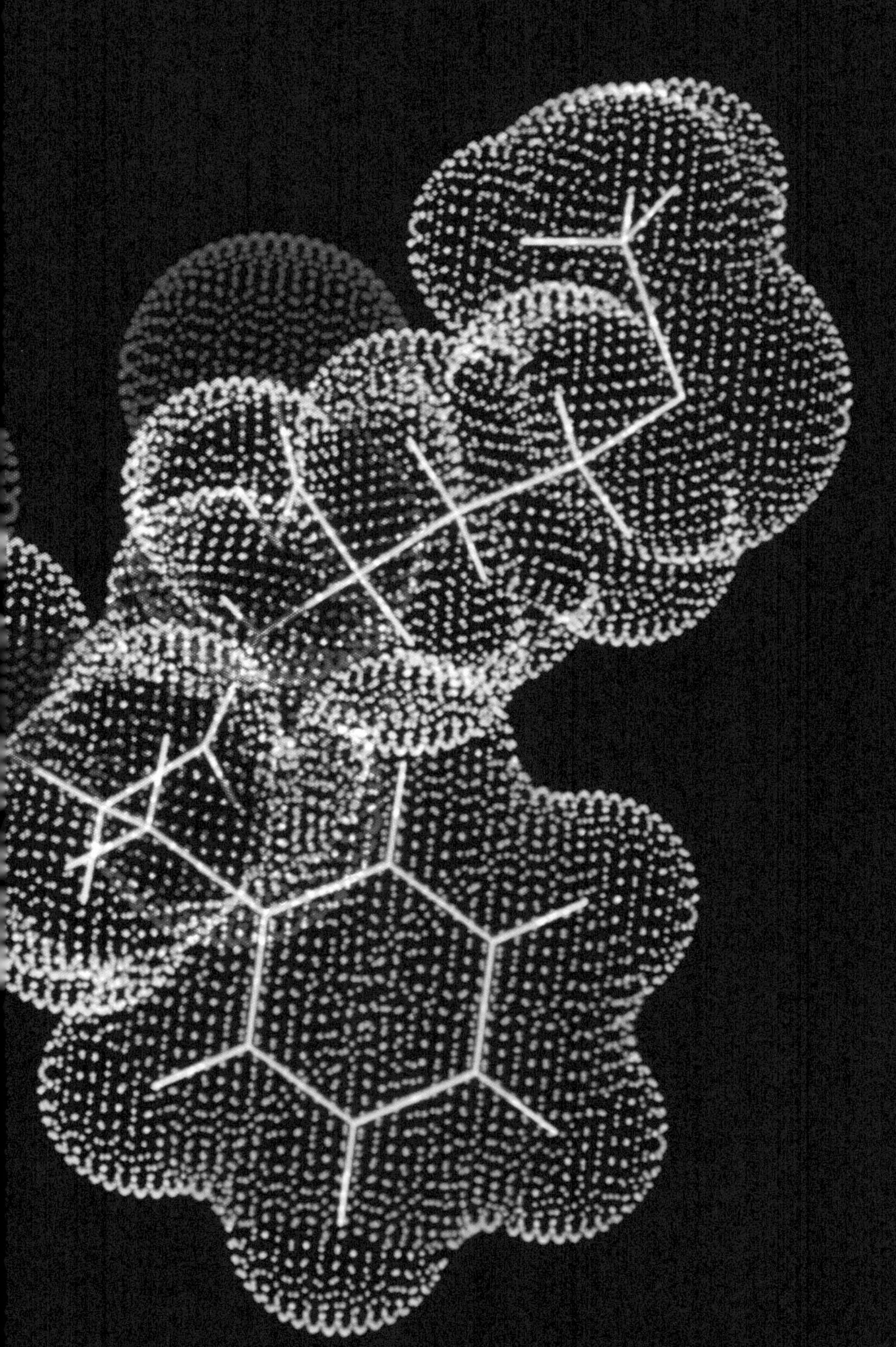

The discovery has made possible the development of non-addictive yet powerful painkillers, transforming the lives of many thousands of people.

It has also revolutionised our understanding of how and why psychoactive drugs affect the brain and provided an insight into why some people become addicted to drugs.

Porter worked out how the armies of antibodies of the immune system attack infections and diseases.

Antibodies, part of a group of proteins in the blood called immunoglobulins, attach themselves to infectious alien organisms and viruses, rendering them inactive.

In 1975 César Milstein of the MRC Laboratory of Molecular Biology and the University of Cambridge and Georges Kohler, based at the Basel Institute in Switzerland, found a way of mass producing pure antibodies called monoclonal antibodies.

The process paved the way for the modern biotechnology industry and the development of new types of drugs and diagnostic tests, in fields as diverse as cancer, the prevention of transplant rejection, pregnancy testing and treatment of arthritis.

Antibodies attach themselves to viruses rendering them inactive

The work underpinned new ways of combating disease through immunisation, where organisms are injected into the body to trigger the production of certain types of antibodies that will repel common diseases, such as tuberculosis, Hepatitis B, polio and smallpox.

Porter's work laid the foundations for a series of medical developments.

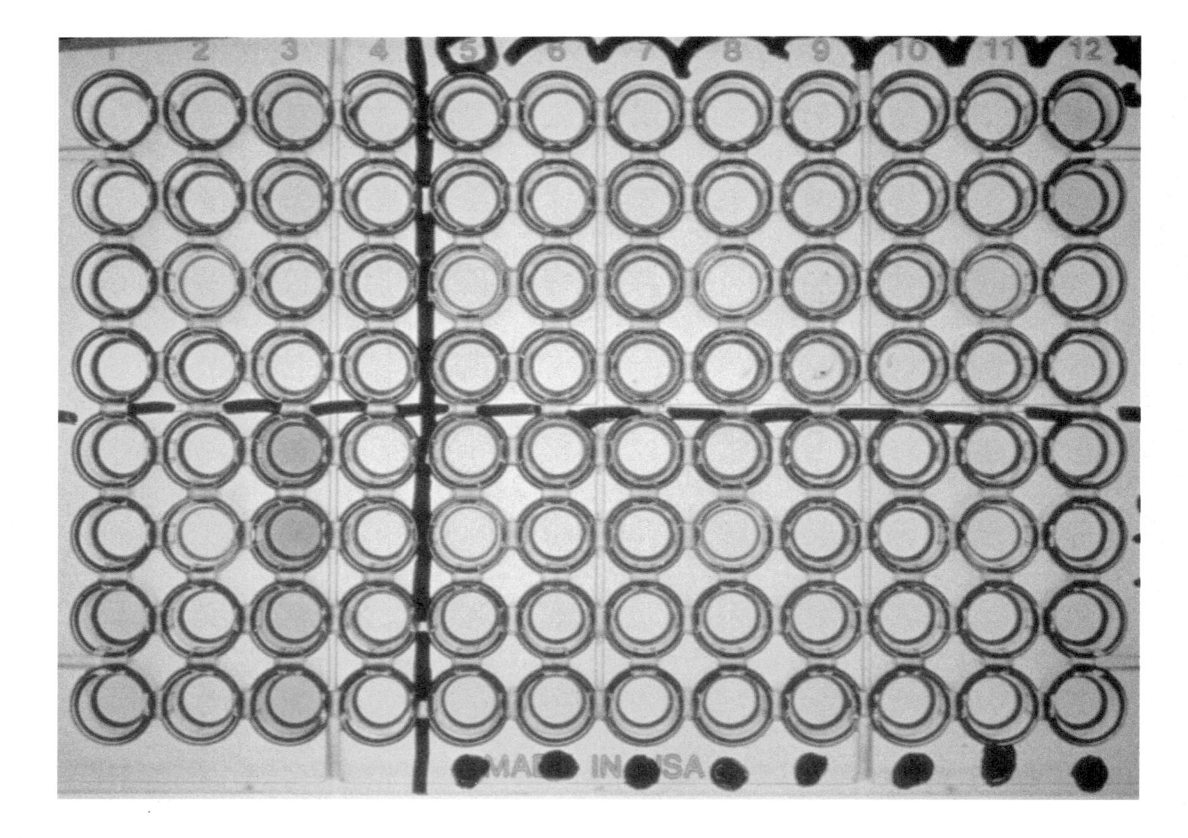

NATURAL DEFENCE

In 1967 Rodney Porter at the University of Oxford helped to uncover the secret to the body's defence mechanisms.

In 1985 Alec Jeffreys at the University of Leicester developed a reliable way to detect differences in the DNA of individuals – a technique now known as genetic fingerprinting.

UNIQUE PATTERNS

UK police now have a database of over two million genetic profiles

Genetic fingerprints are like real fingerprints in that they are unique to every individual (except identical twins). In criminal investigations genetic fingerprints are recoverable from samples at crime scenes, allowing the identification of victims and assailants.

There are now over two million genetic profiles of people in the UK's national police database.

In paternity testing, genetic fingerprinting provides a way of proving fatherhood or other family relationships, essential for example in immigration casework. In medicine, the genetic fingerprints of cells within tumours have aided understanding of how tumours develop. DNA fingerprinting has also been used in criminal investigations of the illegal importation of animals and animal products. It has also illuminated for the first time the breeding strategies of animals as diverse as monkeys and sparrows.

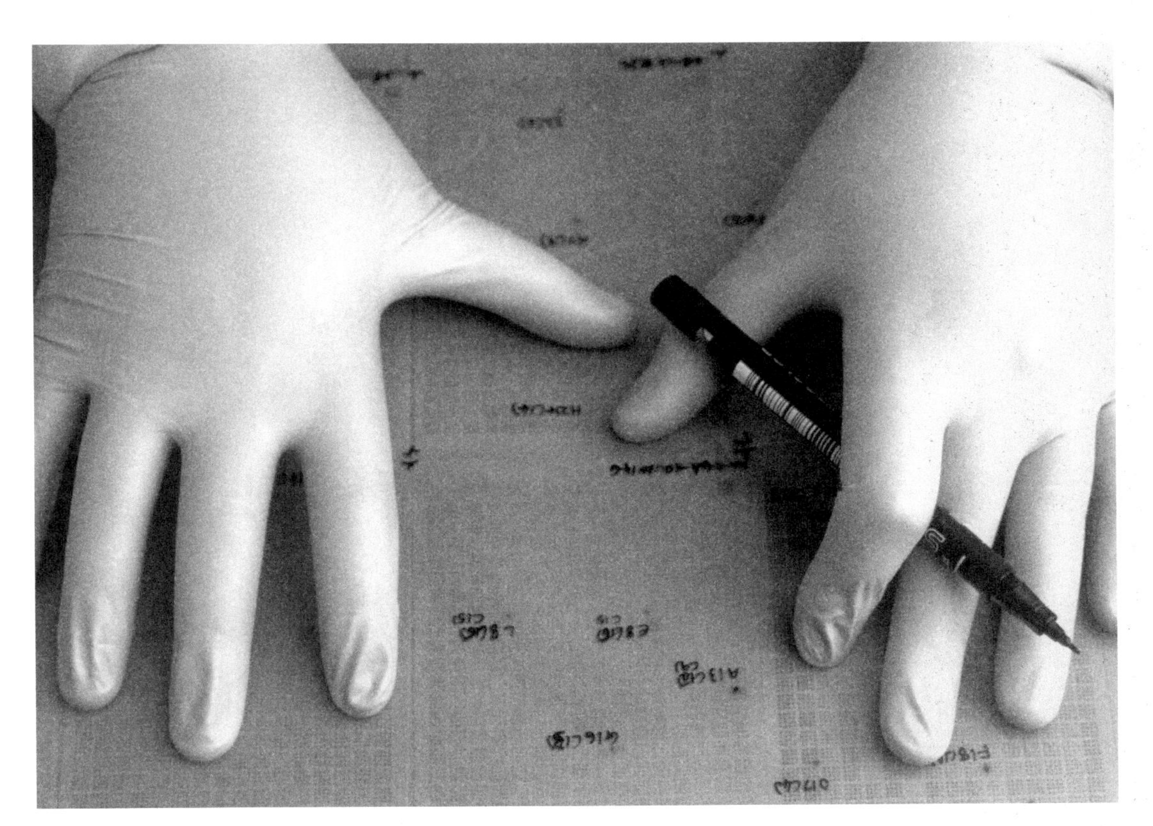

process that regulates the way cells divide and multiply, allowing an egg to grow into an embryo, a baby and finally a full adult.

Every second, millions of cells divide in our body to create new cells as other cells die. First a cell grows in size, then it copies its DNA and splits into two cells called daughter cells. But sometimes the cycle can go out of control, producing millions of unwanted cells. This is called cancer.

Nurse developed much of the work that led to the discovery during his time at Oxford University, while Hunt pursued his early research at the University of Cambridge.

MULTIPLYING HOPE

An adult human being is made up of about 100,000 billion cells; and they all originate from the same single cell, the fertilised egg.

Every second, millions of cells divide in our body to create new cells

The development is one of several made by Nurse and Hunt over the last few decades in their attempts to unravel how the cell is driven from one stage to the next in the cell cycle, and how cancer develops. Mutations in the cell cycle genes may be involved in the development of cancer cells.

The discoveries have opened up new possibilities for treatment of cancer, which remains one of the biggest killers in the world. They will allow doctors to better identify tumours observed in breast cancer and cancer of the brain for example. In the future it is hoped that the work will underpin the development of more effective cancer drugs.

DOLLY AND IAN

In 1997 Dolly the sheep, the world's first animal cloned from a cell taken from an adult animal, was introduced to the world by Ian Wilmut, a scientist at the Roslin Institute, an associated institution of the University of Edinburgh.

Until then, almost all biologists thought that the cells in our bodies had fixed roles. The creation of Dolly from a mammary gland cell of a six year-old sheep showed this was not in fact the case.

Dolly had been created by taking a nucleus from a cell of a mature ewe (Dolly's parent) and inserting it into a hollowed-out donor egg. By jump-starting the egg with electricity, an embryo was created. It then gestated in a surrogate mother to give rise to the first clone using this method of 'cell nuclear replacement'.

Cloning for therapeutic purposes promises many future applications for medicine and agriculture. Herds of similarly cloned animals could one day form the basis for natural 'factories' that produce useful chemicals such as insulin in their milk. Others could house animals that are immune to fatal diseases such as BSE.

Cloning for therapeutic purposes promises many future applications

Another potential development is to produce large animals carrying genetic defects that mimic human illnesses, such as cystic fibrosis. Although mice models for this disease have provided some information, sheep are expected to be more valuable for research, because their lungs resemble those of humans. And as they live for years, scientists can evaluate their long-term responses to treatments.

Some scientists predict that in the future genetically modified animal organs could even be used for transplantation into humans.

Dolly died in 2002 aged seven.

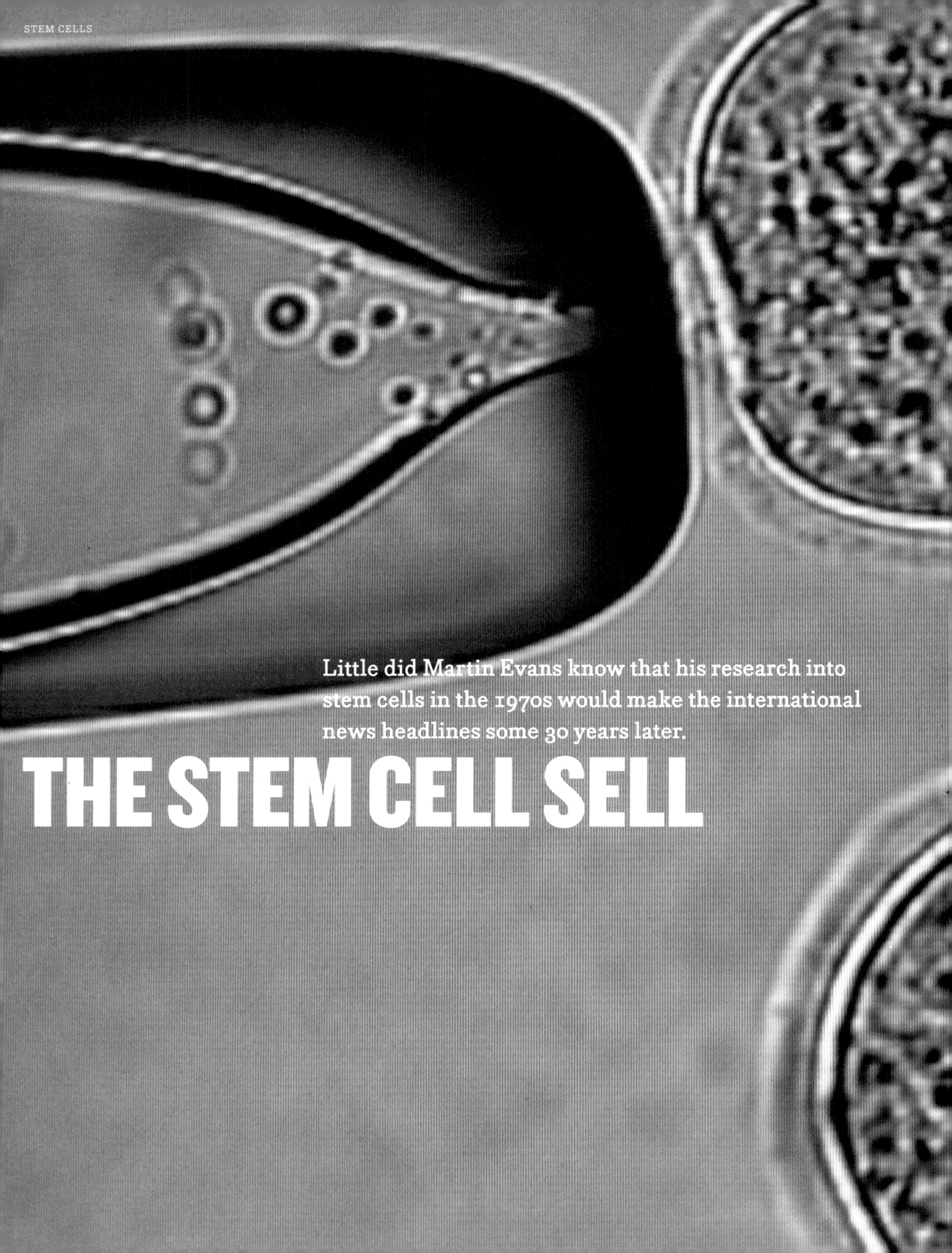

Little did Martin Evans know that his research into stem cells in the 1970s would make the international news headlines some 30 years later.

THE STEM CELL SELL

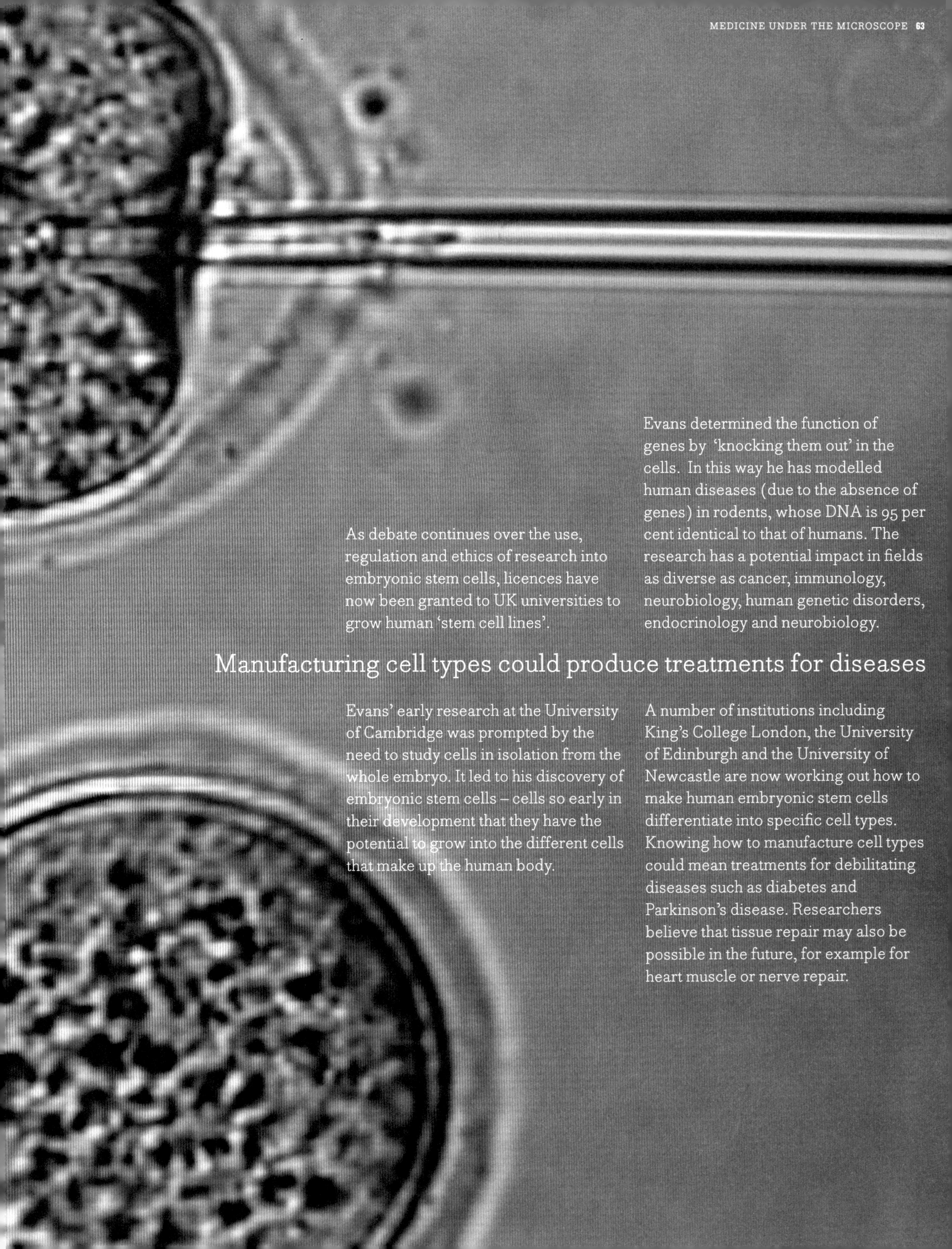

As debate continues over the use, regulation and ethics of research into embryonic stem cells, licences have now been granted to UK universities to grow human 'stem cell lines'.

Evans determined the function of genes by 'knocking them out' in the cells. In this way he has modelled human diseases (due to the absence of genes) in rodents, whose DNA is 95 per cent identical to that of humans. The research has a potential impact in fields as diverse as cancer, immunology, neurobiology, human genetic disorders, endocrinology and neurobiology.

Manufacturing cell types could produce treatments for diseases

Evans' early research at the University of Cambridge was prompted by the need to study cells in isolation from the whole embryo. It led to his discovery of embryonic stem cells – cells so early in their development that they have the potential to grow into the different cells that make up the human body.

A number of institutions including King's College London, the University of Edinburgh and the University of Newcastle are now working out how to make human embryonic stem cells differentiate into specific cell types. Knowing how to manufacture cell types could mean treatments for debilitating diseases such as diabetes and Parkinson's disease. Researchers believe that tissue repair may also be possible in the future, for example for heart muscle or nerve repair.

ROSALIND FRANKLIN
THE WOMAN WHO HELPED REVEAL THE STRUCTURE OF DNA

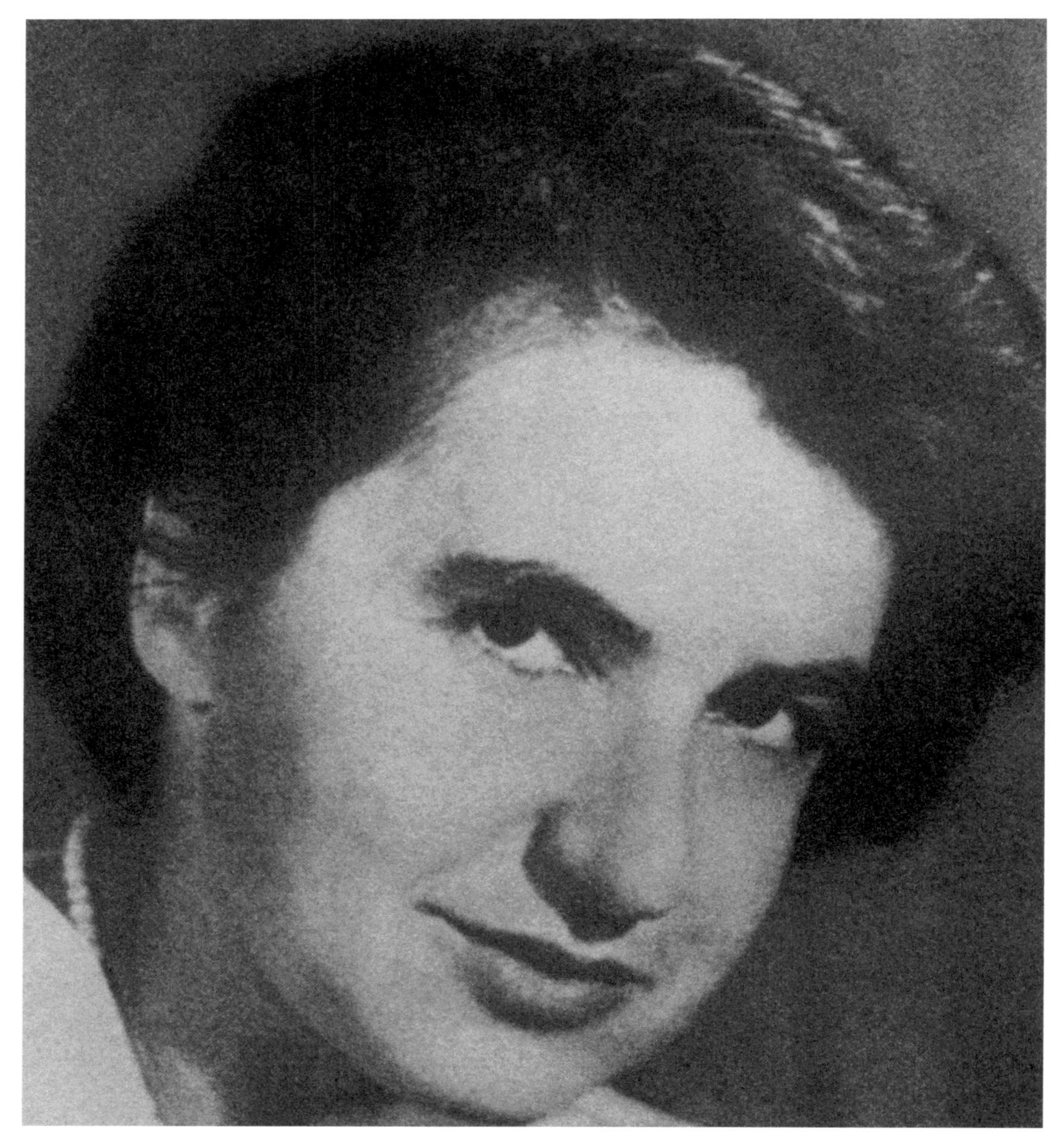

Ground-breaking X-ray images that made the discovery possible

When Francis Crick, Jim Watson and Maurice Wilkins were awarded the Nobel Prize in 1962 for elucidating the structure of DNA, Rosalind Franklin's name was hardly mentioned in their acceptance speeches. It is only now, 50 years on from the discovery of 'the secret of life', that the woman once referred to as the 'dark lady' of the laboratory has come out of the shadows to receive proper recognition for her ground-breaking X-ray images that made the discovery possible.

Franklin died from ovarian cancer at the age of 37, four years before Crick, Watson and Wilkins, Franklin's supervisor at King's College London, were awarded the greatest prize for science. It is thought that the cancer could have been caused by over-exposure to radiation during the course of her research.

Franklin's treatment as a female scientist during a time when men dominated university laboratories continues to provoke controversy to this day. Many argue that Watson stereotyped and belittled Franklin in his classic 1968 book, *The Double Helix*, heralded as one of the most frank accounts of how scientists compete against each other in the race to make a discovery.

Franklin, a University of Cambridge graduate, arrived at King's College to work with Wilkins in 1951 after three years of X-ray diffraction work in Paris. Franklin's task was to utilise X-ray images to map the tiny molecular structure of DNA. But relations between Franklin and Wilkins quickly soured, and soon they were not even on talking terms. He initially mistook her for an assistant rather than an equal, and thought she was too secretive with her research project.

Controversially Wilkins showed the X-ray photos to Crick and Watson without her knowledge or permission, arguing that they belonged to the lab and the institution. The images provided a crucial confirmation of some of the ideas that the Cambridge scientists were developing at the time. Max Perutz, another Cambridge scientist, meanwhile passed Crick and Watson a confidential report, including Franklin's detailed notes and X-ray photographs, received as part of a Medical Research Council appraisal of work at the King's College laboratory.

Watson later revealed how pivotal Franklin's unpublished X-ray diffraction patterns were in confirming the helical structure of DNA, revealed to the world in a joint paper by Crick, Watson and Wilkins in the science journal *Nature* in 1953.

After her DNA work, Franklin took up a position at Birkbeck College, where she worked on the tobacco mosaic virus, which was wiping out tobacco plants. She had begun work on the polio virus when she was diagnosed with cancer. It remains unclear why Franklin did not receive the Nobel Prize posthumously in 1962. But there now exists a Rosalind Franklin prize, awarded to successful female scientists in the UK.

King's College London honoured its DNA pioneers in the naming of its Franklin Wilkins building, opened in 2003.

DISCOVERIES FOR THE DIGITAL AGE

Make an overseas phone call, surf the Internet on your computer, play a CD, and you will have benefited from the work of UK academics. The inventions and discoveries that make possible the electrical gadgets of the modern silicon age follow a long UK tradition: Michael Faraday uncovered the laws of electro-magnetic induction; James Maxwell developed the governing equations for electro-magnetic phenomena; and Joseph John (JJ) Thomson discovered the electron.

UK researchers have been instrumental in developing optical fibres, which now criss-cross the globe, carrying the millions of signals from telephones, cable TV and the Internet. They have also produced the Liquid Crystal Displays that make the flat screens of computers, mobile phones and digital music players possible. They have helped to develop the lasers that underpin CD and DVD players, and discovered how to create holograms, the three-dimensional images seen on bank notes.

Developments promised for the future meanwhile include the world's first hand-held 3D laser scanner that can instantly translate physical objects into three-dimensional models on the computer. Academics are also developing electronic and robotic gadgetry to transform the lives of the disabled.

The world's first computer program was run not in the US, but by University of Manchester scientists, creating what became the prototype for the first commercial computer. It was a Cambridge academic and his research student, meanwhile, who invented the Scanning Electron Microscope, a key tool for the modern microelectronics industry.

Often it is simple curiosity and the buzz of discovery that drives scientists to explore theories and phenomena. Practical applications emerge later. The Cambridge physicist Nevill Mott was interested in fundamental questions such as 'Why is glass transparent?' Thanks to his work, we now have more affordable personal computers, pocket calculators, and photocopying machines. Harry Kroto, meanwhile, jointly discovered the third form of carbon on the planet known as 'Buckyballs'. These new molecules may one day give rise to the next generation of semi-conductors, longer life batteries, or even anti-HIV therapies.

SEEING THE LIGHT

Communication across the world – from telephones to cable TV and the Internet – is made possible by millions of kilometres of optical fibres criss-crossing the globe.

At a cost of a few pennies, we can now speak to someone on the other side of the world with the same clarity as if talking to a next-door neighbour. Huge amounts of sounds, images, video and data can be sent millions of kilometres, at the speed of light, with virtually no distortions or errors.

All this would not have happened if it were not for the research work of a few academics working in UK universities.

In the 1950s the 'founding fathers of fibre optics' Narinder Kapany and Harold Hopkins at Imperial College London demonstrated that light could bend, given the right encouragement. In 1954, Hopkins invented the flexible endoscope enabling doctors to see inside patients' bodies.

Fibre optics carry 1,000 times more information than copper wires

At the University of Southampton during the 1960s Alec Gambling was the first to develop threads of glass, or fibres, suitable for long-distance transmission of light. Fibre optic cables can transmit one thousand times the information that copper wires carry, and are unaffected by electromagnetic interference, so that communication is much clearer and reliable. The invention of lasers at around the same time provided the source of light for the fibres.

The world market for fibre optics is currently estimated to be about $20 billion, while the market for optical communications is about $60 billion. Both are forecast to grow at around 25% per year over the next five years, driven by the expansion of the Internet, company intranets and telecommunications more generally.

QUANTUM LEAP

The Internet, CDs and DVDs have all been made possible through a technology called strained quantum-well lasers that was first proposed by Alf Adams at the University of Surrey.

The technology allows lasers to read and write data on to CDs and DVDs

Quantum-well lasers work by translating information into pulses of light, or photons. The quantum-well is a very thin layer of semiconductor similar in size to the wavelength of an electron. Adams showed that introducing strain to compress the atoms in the quantum-well made the lasers able to handle larger quantities of information with much greater efficiency and accuracy.

It enables computers to send information along fibre optic cables across the globe. It is also used to read and write data using lasers inside almost every CD and DVD player. Strained quantum-well lasers are also used in laser printers and in scanners used to read print. As a result, almost all semiconductor lasers produced commercially (about one billion per year) now contain strained quantum-wells, in an industry worth some £10 billion.

Many electrical gadgets that use flat screens – computers, mobile phones, pocket calculators, digital watches, and digital thermometers – depend on liquid crystals.

LIQUID GOLD

Liquid crystals are materials that can interact with light in different ways when responding to small changes in external stimuli such as small electric voltages. When placed between transparent electrodes they can be used to create pixels in Liquid Crystal Displays (LCDs), which combine low weight and slimness with low power consumption and low operating voltages.

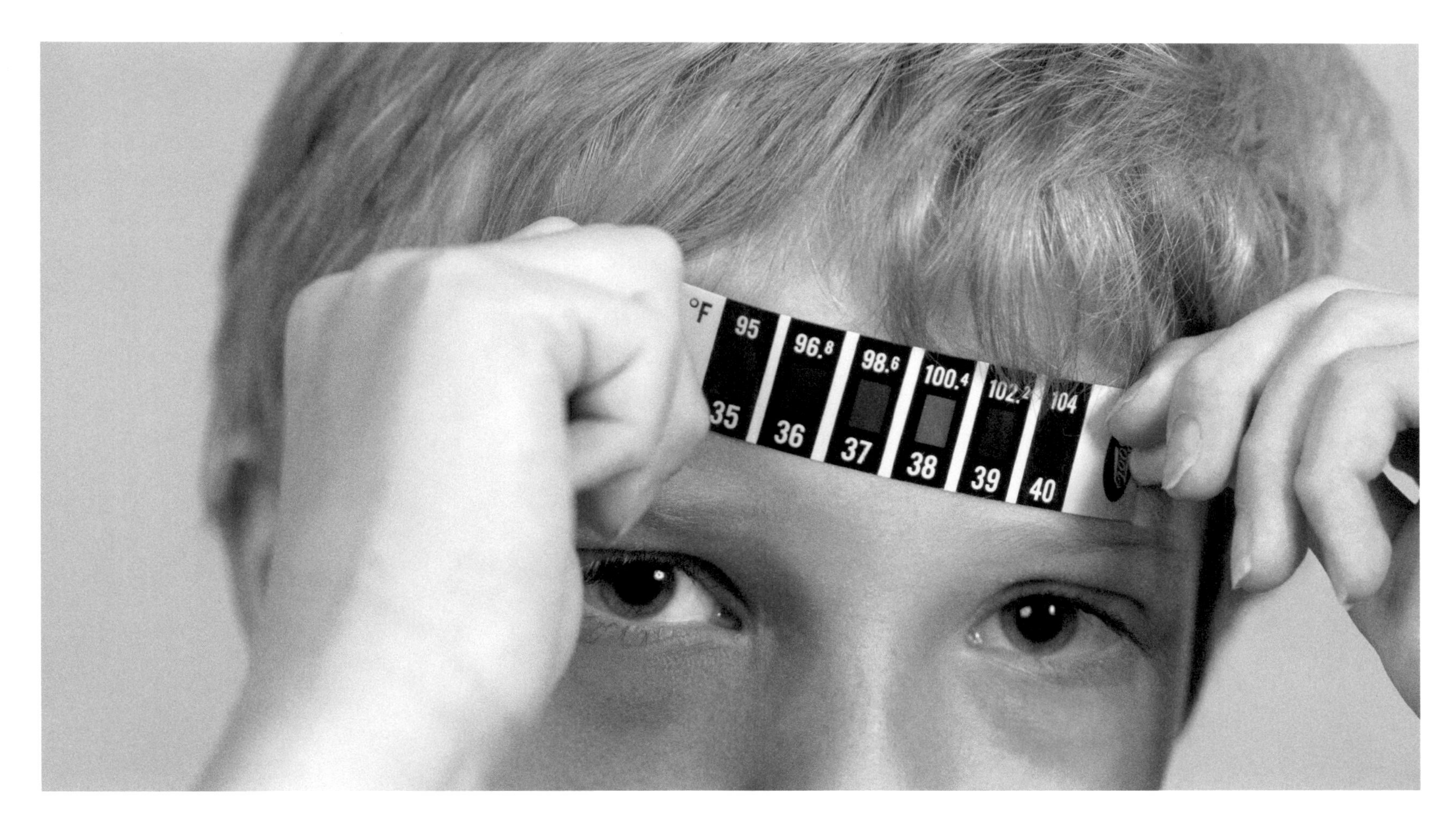

Liquid crystals are the basis of a billion dollar worldwide industry

The first stable liquid crystals for use in LCDs were created by George Gray and his colleagues at the University of Hull in the early 1970s. Before Gray's research, liquid crystal materials were not very stable at room temperature when exposed to moisture, oxygen or light, and could not be handled very well.

It was Gray who discovered a new class of liquid crystalline materials that were stable. The twisted nematic LCD invented in 1972 is still used to this day – for example in computer screens.

The communications industry has exploited this technology by steadily increasing the size and enhancing the performance of LCDs, first in small digital watches, then in calculators, and finally in computer and even TV screens. Healthcare has also benefited with the use of heat sensitive liquid crystals in non-invasive thermometers to monitor pregnancies and identify certain types of breast cancer.

Liquid crystals are the basis of a billion dollar worldwide industry, estimated to be of $40 billion in 2005.

The word hologram comes from the Greek words 'holos', meaning whole and 'gamma', meaning 'message'.

WHOLE MESSAGE

It is used in bank notes, toys and on music products

A hologram records three-dimensional images on two-dimensional surfaces, and allows viewers to see different perspectives of an object, as if it were actually there.

Dennis Gabor at Imperial College London, invented the method of producing holograms. It was a spin-off from attempts to improve the resolution of the scanning electron microscope, which at the time was not powerful enough to properly map atomic lattices. Gabor's classic holograms are still reproduced today.

Two University of Manchester scientists, Freddie Williams and Tom Kilburn, are credited with developing the world's first stored-program computer. In the late 1940s and early 1950s they produced a series of breakthroughs in the early development of computers.

MANCHESTER MEMORY

The work at Manchester came out of the radar experience of Williams and Kilburn during World War Two. They had the idea that a cathode ray tube could be developed to store and make available information ('memory') at the same speed as electronic circuits. The memory system they invented, later called the Williams Tube Memory, was subsequently used in many commercial computers, including IBM's.

They developed the first modern computer that functioned as a complete system – the 'Manchester Mark 1'. It became the prototype for the first commercial computer, the 'Ferranti 1'.

Williams and Kilburn went on to build the world's first computer with transistors, and the world's first programmable computer, 'Atlas'. At the time, it was the most powerful and sophisticated computer in the world, setting the benchmark for subsequent machines.

Willliams and Kilburn developed the first modern computer system

MICRO DESIGN

The Scanning Electron Microscope (SEM) allows researchers to peek inside materials, right down to the level of their most basic building blocks, atoms, enabling them to design materials that have the right properties to fit many different purposes.

Engineers have even designed some electron microscopes to build materials atom by atom. Engineering on such a small scale is crucial to microelectronics, where huge complex electronic circuits are packed into spaces no bigger than a pinhead.

The microscopic designs have underpinned space travel, advanced surgical techniques and the basic personal computer. Behind the scenes, SEMs have also changed crime and accident investigations, revealing tell-tale marks that provide information about the individual gun that fired a bullet, or the manner in which a jet engine failed.

SEMs are now used to create today's electronic chips

The microscope was developed by Charles Oatley and his research student, Dennis McMullan, at the University of Cambridge in the 1950s.

The microscope had been rejected by research groups in America and elsewhere, but the two researchers showed the device was in fact practical and had many potential advantages.

Commercial interest was at first slow. But SEMs are now used to create today's electronic chips, with about 100,000 currently in use worldwide.

'ModelMaker' is the world's first hand-held 3D laser scanner that can accurately and quickly scan physical objects to make colour three dimensional computer models.

PAINTING IN 3D

The novel technology has been pioneered thanks to the work of researchers at the University of Surrey, who have developed computer software, which allows surface modelling and versatile fast 3D scanning of complex objects to transform optical surface measurements into 3D models.

Applications include: computer animation for films; product design and styling; virtual museums and restoration; 3D catalogues on websites; and capturing human geometry for robotic surgery.

Objects ranging in size from millimetres to meters can be scanned

Scanning with the device is similar to painting with a 50mm wide paintbrush. While scanning, the user receives immediate feedback from an image of the object's scanned surface appearing in real-time on the computer. Objects ranging in size from millimetres to metres can be scanned.

HANDY HELPER

Even the most basic tasks such as shaving, cooking or cleaning can be a struggle if you are disabled or elderly, forcing you to rely on others just to get through life's daily chores. But a robotic caterpillar developed by scientists at Staffordshire University allows people to perform the basic tasks of day-to-day living by themselves, with the privacy and dignity that the able-bodied take for granted.

'Flexibot' is the world's first robotic arm that can propel itself like an inchworm from one socket to another across a room. It takes the form of an arm, jointed in the middle and at either end, which can clamp itself to sockets on a wall or ceiling. By plugging one end into a socket and then reaching over and plugging into the next one, it can move around on its own accord. Taking instructions from each 'intelligent' socket, it performs different tasks using its three-fingered hand to grip and manoeuvre objects.

Several multi-national companies are currently discussing how the robot could be mass-manufactured to transform the lives of people with disabilities across the world. Its benefits are simple: it is cheap, accurate, and flexible. It also works: Flexibot is based on the same principles as 'Handy one', currently the most successful rehabilitation robot in the world.

The robot could transform the lives of people with disabilities across the world

Sheffield Hallam University meanwhile have developed artificial arms that work and move like real limbs. The results of the 'Analogous Artificial Arm' project are already being used by the National Aeronautics and Space Administration (NASA) to help develop a robot skeleton with plastic muscles for future space missions. Other future applications include elbow implants, and machines controlled by computers that can be used to mimic surgical operations.

Researchers at Leeds Metropolitan University are using computers to help people with autism and Asperger's Syndrome learn how to deal with potentially problematic social situations. The simulations ask computer users for example to choose where they should join a queue and illustrate the consequences of their choices. The software is being designed to make it easy for parents, teachers or carers to write new simulations aimed at their young people with autism.

The work was prompted by the quest for the greater theoretical understanding of physical matter, but it has since contributed to a series of practical everyday applications.

Thanks to Mott, we now have more affordable personal computers, pocket calculators, photocopiers, solar panels and other electronic devices.

Working at the University of Bristol and later at the University of Cambridge, Mott's studies explored the electrical conductivity of amorphous materials, so called because their atomic structures are irregular or unstructured.

Why is glass see-through? This simple but fundamental question was one that inspired Nevill Mott to pursue research into how materials conduct electricity and absorb light.

AMOUR AMORPHOUS

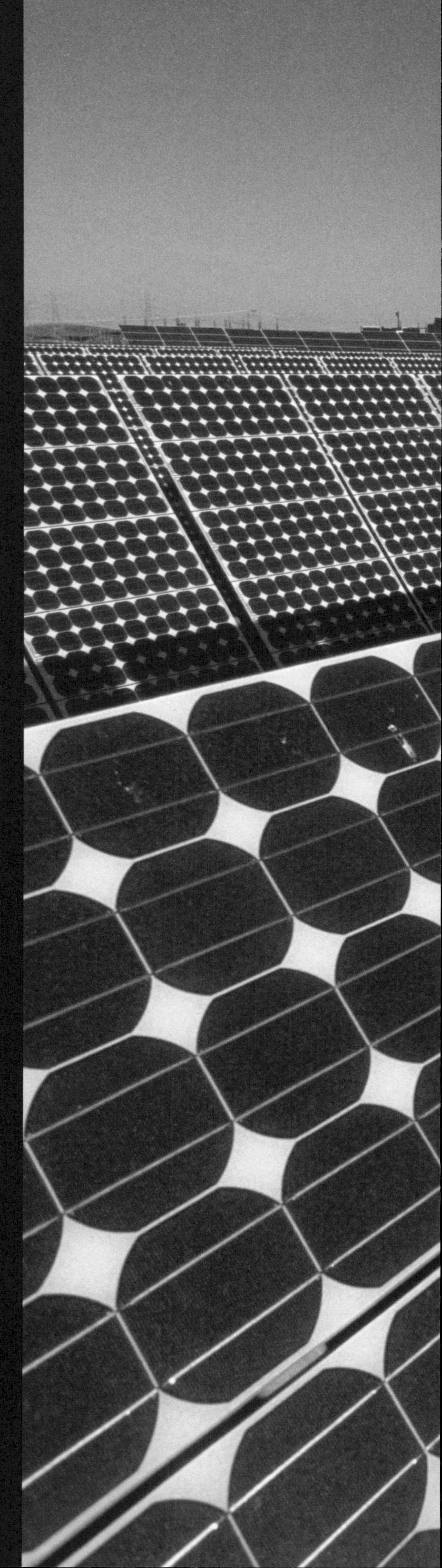

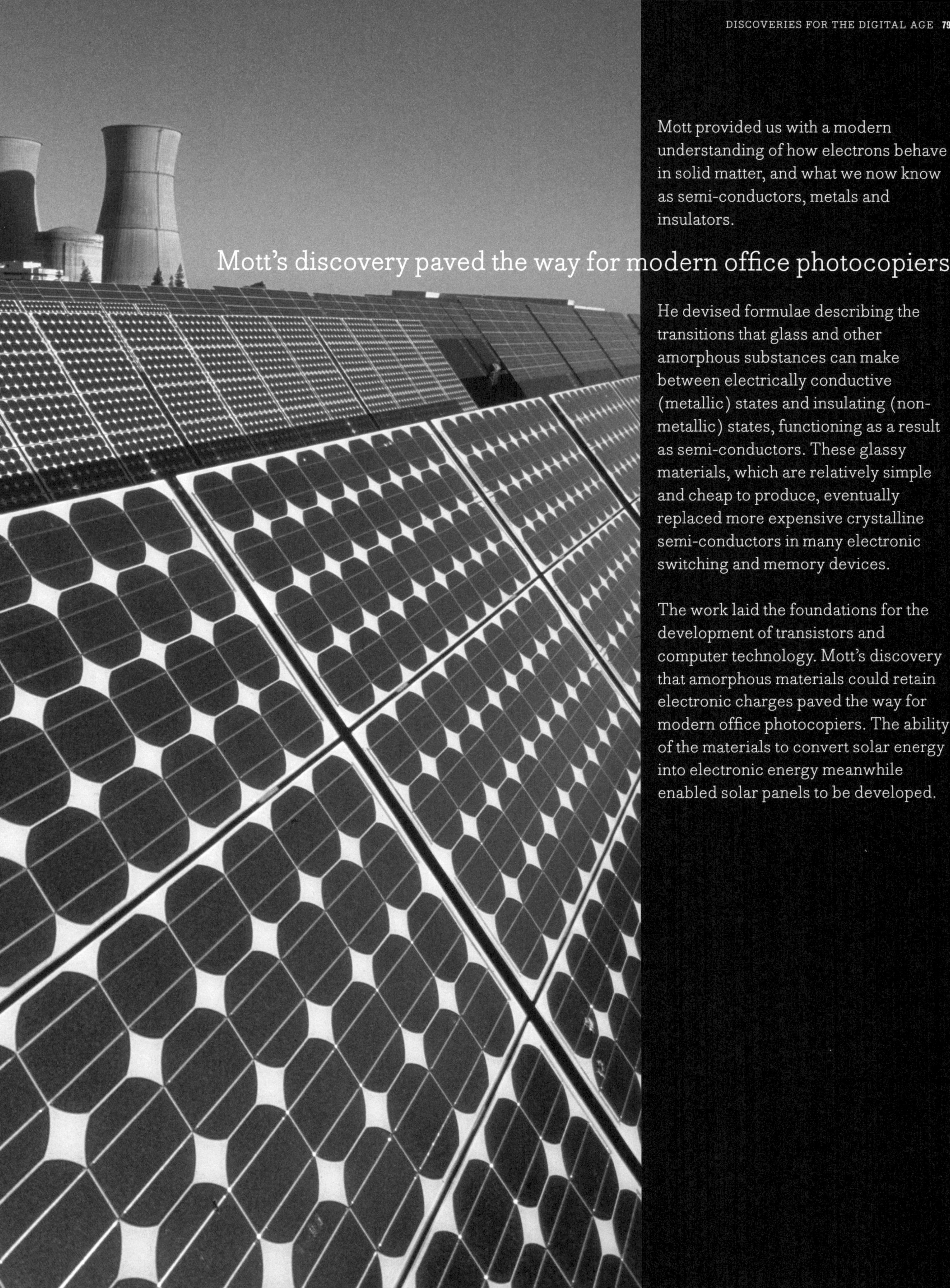

Mott provided us with a modern understanding of how electrons behave in solid matter, and what we now know as semi-conductors, metals and insulators.

Mott's discovery paved the way for modern office photocopiers

He devised formulae describing the transitions that glass and other amorphous substances can make between electrically conductive (metallic) states and insulating (non-metallic) states, functioning as a result as semi-conductors. These glassy materials, which are relatively simple and cheap to produce, eventually replaced more expensive crystalline semi-conductors in many electronic switching and memory devices.

The work laid the foundations for the development of transistors and computer technology. Mott's discovery that amorphous materials could retain electronic charges paved the way for modern office photocopiers. The ability of the materials to convert solar energy into electronic energy meanwhile enabled solar panels to be developed.

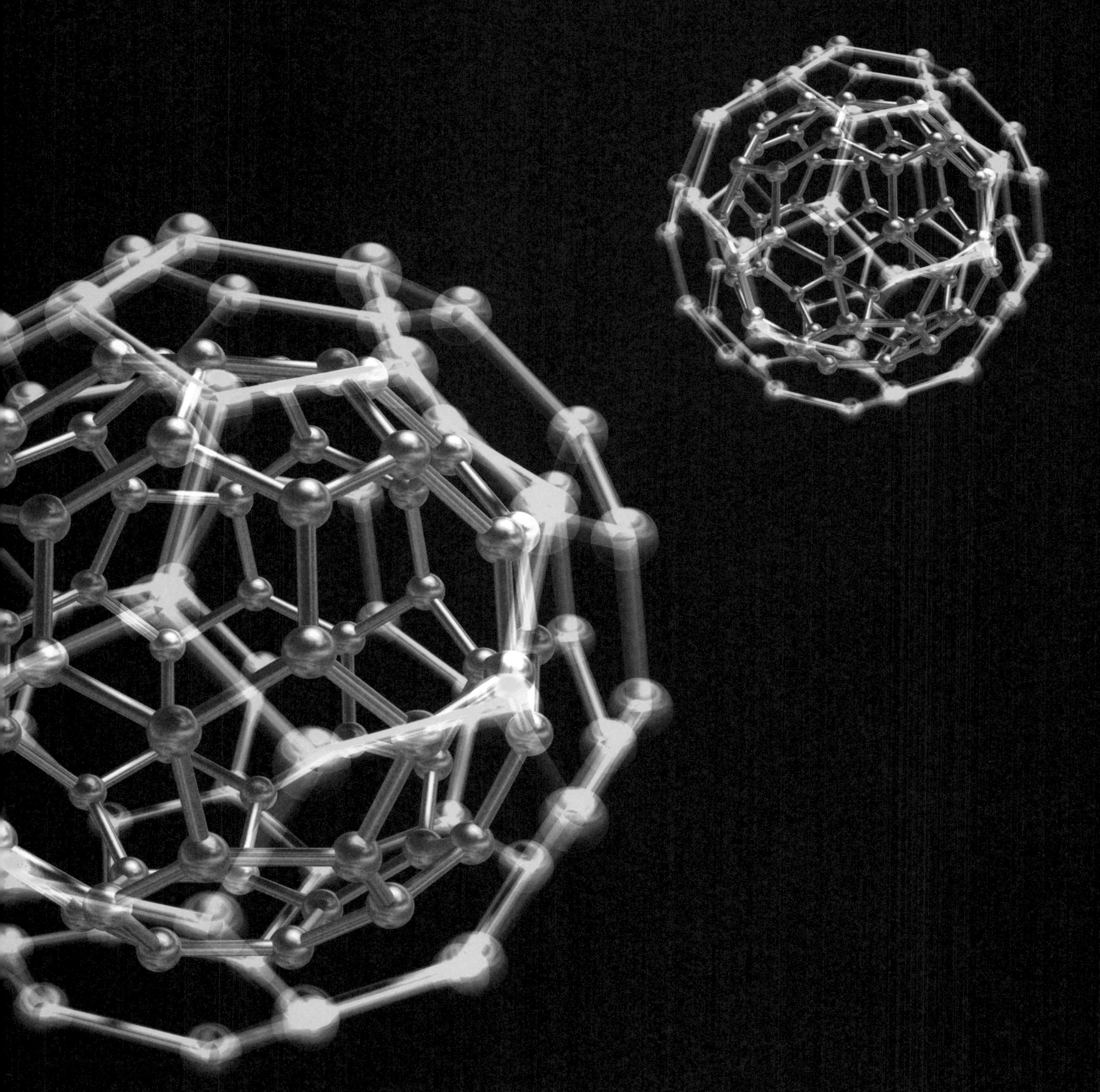

CARBON FOOTY

Carbon forms the basis for most known chemical substances in the world, including those that underpin life, such as DNA and proteins. But it was only in 1985 that the third well-defined form of the pure element was discovered.

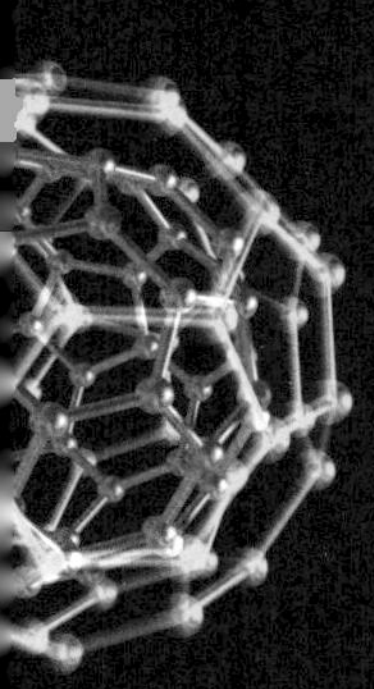

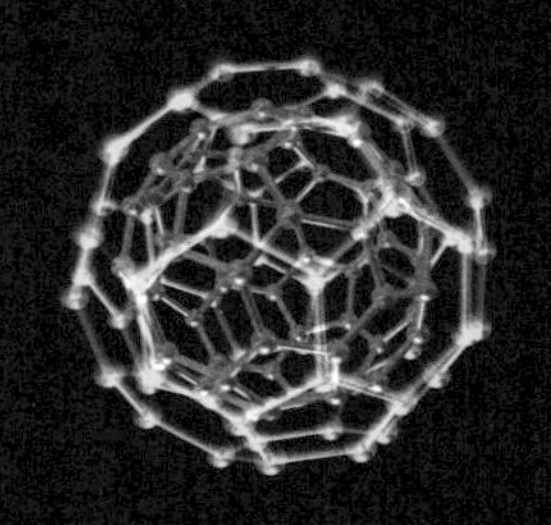

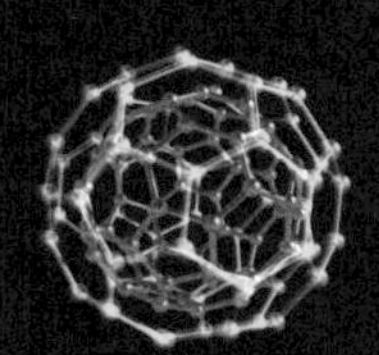

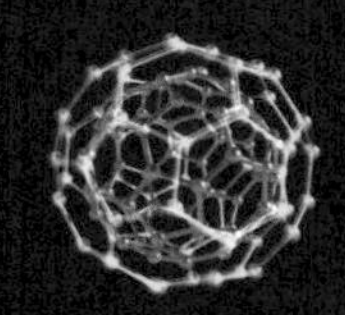

Kroto was interested in long chains of carbon atoms that could be detected billions of kilometres away in space by radiotelescopes.

Scientists have since been able synthesize large quantities of buckyballs, although it is now known that fullerenes are created naturally – every time a candle or an oil lamp burns.

Commercial applications of fullerenes include anticancer therapies

Harry Kroto at the University of Sussex, and his US collaborators, revealed that carbon can exist as tiny spherical molecules, now known as fullerenes or 'buckyballs'. Up until then, scientists knew of only two forms – diamond and graphite.

He and US researchers vaporized graphite with a powerful laser in an atmosphere of helium gas to mimic the high-temperature conditions found near red giant stars in galaxies. They created a number of previously unknown carbon cage molecules, the most common of which contained 60 carbon atoms.

They found that the molecule was like a microscopic football, with carbon atoms linked in a spherical match work of 20 hexagons and 12 pentagons.

The new molecule was named buckminsterfullerene as it resembled the geodesic dome designed by the architect, Richard Buckminster Fuller, for the 1967 Montreal World Exhibition. Chemists write it as C60.

Thousands of patents already exist for commercial applications of fullerenes, including superconductive materials, next generation semi-conductors, longer life batteries, 'artificial diamonds', as well as anticancer and anti-HIV therapies, and new drug delivery systems.

HAROLD HORACE HOPKINS AND NARINDER SINGH KAPANY
THE FOUNDERS OF FIBRE OPTICS

Hopkins and Kapany were able to make bundles of up to 20,000 'optical' fibres and were able to transmit the letters 'GLAS' through their fibre cable.

The founding of fibre optics still prompts contentious debate about which academics should be credited most for its development. Two separate papers were published simultaneously in the science journal *Nature* in 1954. Both detailed the first flexible tubes that enabled light to be reflected through channels – promising a new powerful way of transmitting information. And the joint academic authors of one of the publications were never to agree over which had done most to find a reliable way of transmitting images via light waves.

Few people in the wider world will know the names of these academic pioneers. They were Harold Horace Hopkins and Narinder Singh Kapany, then based at Imperial College London. They were responsible for a technology that went on to transform medical operations, was used on the Apollo spacecraft, and still to this day holds potential to improve the way that information is transmitted across the world. Credit is also given to the author of the other *Nature* paper, Abraham van Heel, Professor of Physics at the Technical University of Delft, in the Netherlands. Hopkins had designed a zoom lens used for BBC TV cameras. He was inspired to pursue his fibre optics work after a dinner party where a doctor had asked whether a flexible instrument could be made to look in to the stomach. (This was later to be known as the endoscope.)

In 1952, Hopkins secured a grant from the Royal Society which included money for a research assistant. Hopkins offered the position to an Indian student named Narinder Kapany.

Born in the Punjab, Kapany joined Imperial to study for a PhD in optics. Legend has it that when growing up a teacher told Kapany that light could travel only in a straight line – inspiring him to find a way of disproving this assumption.

Hopkins and Kapany were able to make bundles of up to 20,000 'optical' fibres and were able to transmit the letters 'GLAS' through their fibre cable.

In November 1953, they sent a letter to *Nature* which got published in 1954 just below van Heel's paper. Ever since there have been accusations and counter-accusations that the rival academic groups had been tipped off about the other's work. Yet till his death Hopkins denied that he had seen van Heel's paper.

In 1955, Kapany obtained his PhD at Imperial and then left to pursue an academic career in the United States. It is said that Hopkins and Kapany never resolved their differences over their respective contributions to the work.

Hopkins died in October 1994. But his name lives on. In 2005 the Hopkins ward opened at the Royal Berkshire Hospital – named after his later work developing the endoscope while at the University of Reading.

In the 1960s Kapany wrote a series of papers that further developed the technology and was the first to coin the phrase 'fibre optics'. This has led to him being known as the father of fibre optics. Kapany also became one of the early academic entrepreneurs in California's Silicon Valley, founding successful optics businesses. And in 1999 Kapany was listed, by the US based *Fortune* magazine, as one of seven unsung heroes who had greatly impacted on life in the 20th century.

PLANES, TRAINS AND AUTOMOBILES

Make a journey by plane, train or automobile, and without knowing it, you will have benefited from the work of researchers in UK universities. They have helped to make modern travel quicker, safer, and more convenient.

Modern design and engineering of all vehicles is based on a technique called 'finite element analysis' developed in part by UK academics. Atomic theory meanwhile has helped to create tougher materials for jet engines. A new field called 'aeroacoustics' has enabled engineers to soften the noise of jet engines.

One day we may even get to travel on the first magnetically levitating trains – invented by the late Eric Laithwaite. Designs of vehicles for films set 300 years into the future meanwhile have inspired a new environmentally friendly car.

And just imagine how unreliable public transport would be without computerised train and bus schedules, developed by academics at the University of Leeds. Researchers also designed the signs that guide you on roads and motorways in the UK.

Finally, researchers have improved the safety of passengers. Aircraft interiors and flight procedures are designed differently today, thanks to work in universities. A revolutionary cooling system has been designed for the London Underground system. And new techniques mean that the country's ageing road bridges are now safer. Car drivers will also for the first time be able to locate where an emergency siren is coming from, thanks to the application of directional sound, an idea that came to an academic while sitting in a traffic jam.

TESTING TIMES

The design and engineering of vehicles and buildings has been revolutionised by a technique called 'finite element analysis' (FEA), which was developed in part by UK academics.

Engineers can predict responses to stress, vibrations and heat

The technique breaks down a complex structure into small elements. A computer program then analyses the forces acting on each element separately, and co-ordinates them to give an overall picture of what happens to the whole structure. Engineers can accurately predict responses to stress, vibrations and heat.

In the 1960s Olgierd Cecil Zienkiewicz at the University of Wales, Swansea developed the finite element method, which allows engineers to test out engine designs in real life simulations.

In the 1970s Bill Dawes and John Denton at the University of Cambridge meanwhile led the field of computational fluid dynamics, which enables researchers to test the dynamics of new designs.

Few of the manufacturing and design techniques needed to make complex automotive parts could exist without these engineering techniques. They were used for example in the design of some Airbus planes, and in 'Thrust', the car that broke the sound barrier in 1997.

Engineers use FEA to test the safety of new vehicle designs in dangerous situations, simulating car, train and ship collisions for example, and the penetration of body armour.

SHOWING THE WAY

The unique road signs that we see in the UK are all thanks to the work of Jock Kinneir and Margaret Calvert at the Royal College of Art.

The signage system has been dubbed the 'corporate identity of Britain'

In the 1960s the researchers created a signage system based around a sans serif typeface, a hierarchical structure for ordering information and colour coding. It has been dubbed by some as the 'corporate identity of Britain'.

The work was commissioned in 1963 following a Government review of traffic signs. At the time the signs were hard to see and read at normal speeds, were often not effective at night, and did not have a uniform appearance.

A new door had opened in the science of materials. It eventually led to the design of tougher materials for jet engines and other high stress environments.

Hirsch's technique is now a powerful tool in materials science

When a metal is deformed permanently by bending or rolling, it changes its shape by the sliding of atomic planes over one another. This sliding occurs gradually by the movement of atomic scale line defects, called dislocations.

ZOOMING IN

In 1956 Peter Hirsch and his collaborators at the University of Oxford observed for the first time the motion of tiny dislocations in the atomic structure of metals.

Hirsch developed a technique where foils of metals thin enough ($\sim 10^4$ mm thick) to be transparent to high energy electrons, were examined in an electron microscope. The researchers applied the technique to study the nature of defects in materials introduced by metal-working, fatigue, quenching, or irradiation in a nuclear reactor.

The technique has been established as a powerful tool in materials science, and its application has had a profound impact on our understanding of the mechanical properties of crystalline materials which depend on the nature, distribution and interaction of these defects. It has been applied to metals, ceramics, semi-conductors, superconductors and minerals.

The work has contributed to the development of new and improved industrial materials, such as high temperature materials for jet engines, light and strong alloys for load bearing structures, alloys resistant to nuclear irradiation in reactors, and the conditions under which materials can be used safely.

HIGH NOISE

Ever since the jet engine was developed 50 years ago (co-invented by the English aviation engineer, Frank Whittle), there has been a demand for less noisy engines. James Lighthill at the University of Manchester was the first to understand how to minimise sound created in jet engines.

In 1948 Lighthill was asked by the Government to find out if military jets could be used for civilian purposes. The problem was how to make jets quieter and more powerful at the same time. In his 'Eighth Power Law of Jet Noise' Lighthill showed how engine noise increased with greater jet speeds.

This knowledge made quieter engines possible. The work led to lower jet speeds and the large-fan aero engines that propel today's passenger aircraft. It would take hundreds of them to make as much noise as a single aircraft from the early jet fleets.

Lighthill created a new field in fluid mechanics, called 'aeroacoustics', or 'sound generated aerodynamically'.

Research helped bring the noise of Concorde into operational bounds

Professor Shon Ffowes Williams at Imperial College in the 1960s and at the University of Cambridge over the last 30 years, then worked on minimising the noise of high-speed supersonic aircraft. The research helped to bring the noise of Concorde within acceptable operational bounds. The work has also resulted in some of the basic methods for designing quieter supersonic fan and helicopter rotors.

Medicine has also benefited from aeroacoustics. Researchers at Addenbrookes Hospital in Cambridge have applied the same principles to control the breathing instabilities responsible for obstructive sleep apnoea and snoring.

LEVITATION STATION

In the 1950s Eric Laithwaite at Imperial College London designed the world's first magnetically levitating train.

The trains float on a magnetic field, propelled by a 'linear induction motor'. 'MagLevs' have no traditional onboard motor or steel wheels like traditional trains. Instead they are driven forward by magnets that line a guiding track on the ground.

Laithwaite argued that the trains have two main advantages over traditional trains: less maintenance and higher speeds. As the train floats along there is no wear and tear caused by friction. The floating action also means that the MagLevs go fast, reaching speeds of 300mph or 500kmh.

Laithwaite's explanation of the linear induction motor triggered controversy and opposition among researchers as he claimed it defied the known physical laws of motion – a sort of perpetual motion machine.

The UK Government withdrew support for Laithwaite's ideas, but the design was taken up in Japan and Germany where high-speed demonstration MagLev trains have been operating since the 1970s.

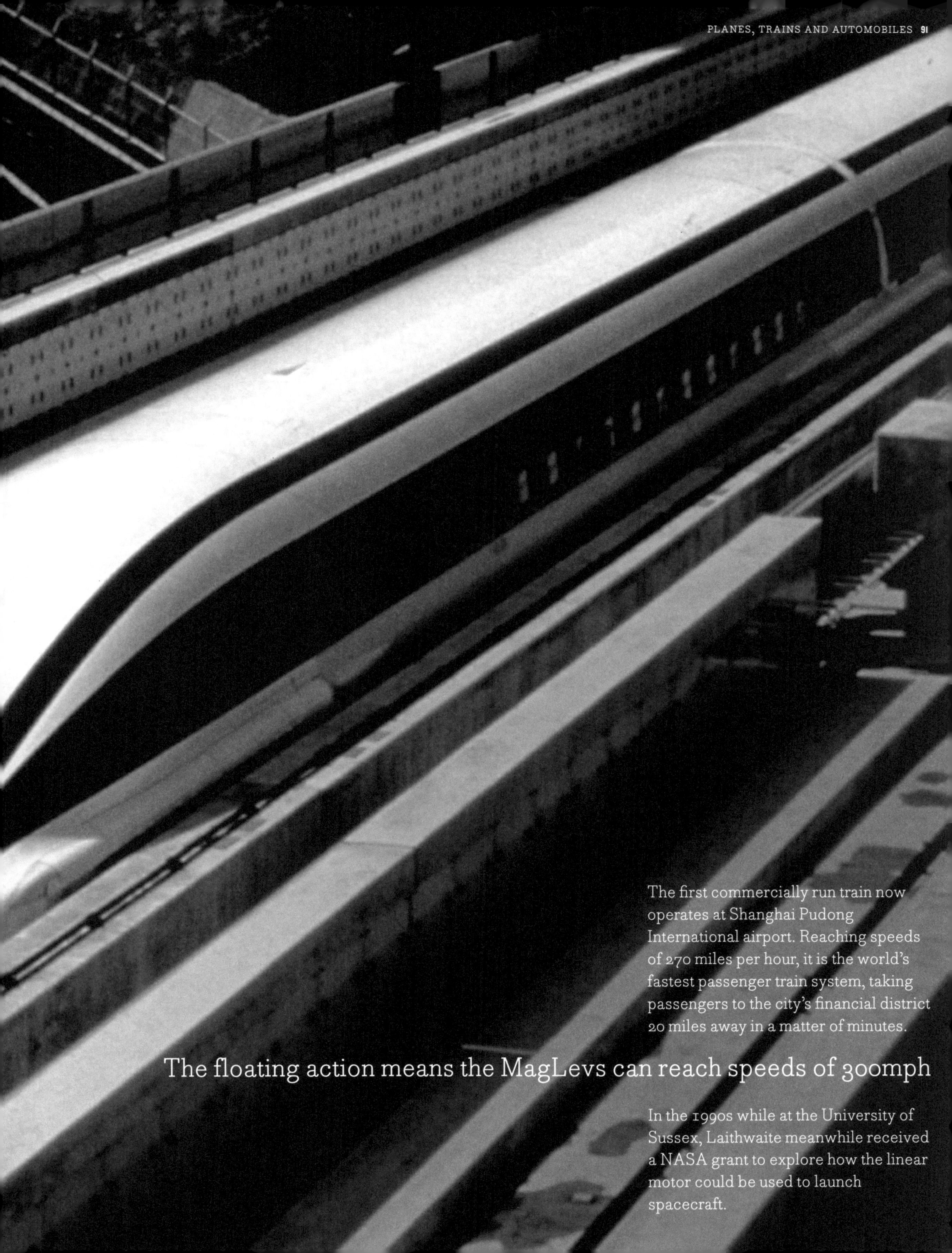

The first commercially run train now operates at Shanghai Pudong International airport. Reaching speeds of 270 miles per hour, it is the world's fastest passenger train system, taking passengers to the city's financial district 20 miles away in a matter of minutes.

The floating action means the MagLevs can reach speeds of 300mph

In the 1990s while at the University of Sussex, Laithwaite meanwhile received a NASA grant to explore how the linear motor could be used to launch spacecraft.

'Microcab' is an environment friendly car inspired by the designs of vehicles for films set 300 years into the future. Not only this: the Coventry University lecturer behind the car also built the robot R2D2 for the *Star Wars* films (which also inspired the design for London's millennium bridge).

BREATHING EASIER

The key idea behind Microcab is that it is powered by technology that does not give rise to the harmful fumes generated by traditional petrol powered engines.

Microcab is one of the few purpose built hydrogen fuel cell vehicles in the world. The fuel cell generates electricity by reverse electrolysis – bringing hydrogen and oxygen together to produce water and electric current that propels the car's electric drive.

Hydrogen and oxygen are brought together to produce water and electricity

The vehicle will have a range of about 110 miles between refills, with a tank of hydrogen probably costing about 70 pence.

The idea for a future urban taxi came to John Jostins when stuck in a traffic jam on the way to a film special effects unit in London. The main impact will be on urban air quality – cities are still heavily polluted by current cars.

The first prototype of Microcab has now been running for some years. Currently there are plans to run a series of pilot schemes around the country to help refine the design before approaching car manufacturers.

LIKE CLOCKWORK

British Rail introduced the world's first computerised train schedule in 1963 – designed by Tony Wren at the University of Leeds. It helped to improve the efficiency of the freight trains serving the London docks.

The system used a computer the size of a small house, but with less power than a modern desktop PC.

Bus, train and driver scheduling are incredibly complex mathematical problems. Wren's system incorporated 'trial and error' into calculations, using computational formulae or procedures known as algorithms. It ensured locomotives and their drivers didn't unnecessarily move without passengers or wait around, and waste resources.

Wren's original concept is now used in transport systems worldwide

Wren's original concept is now in use in transport systems all over the world. The model is very sensitive, able to calculate the impact of, for example, drivers' lunch breaks being just a few minutes longer.

In 1975, Manchester became the first city in the world to use the system to schedule its buses. In London, an advanced version was used in 1985 to schedule drivers for the first time. Related systems have been used by retailers to reduce the numbers of trucks on the road, while increasing deliveries.

Research in computerised scheduling continues at Leeds using new combinations of mathematics and artificial intelligence. The most recent system designed at the University has been installed in the UK's largest bus group and is being widely used by train operators.

Computerised scheduling has saved transport companies money through extra efficiency, and given the public a better and less expensive service.

The research team were the first to reproduce the real human behaviour that occurs in aircraft emergencies – the smoke-filled cabin and resulting panic that will occur.

They built an aircraft simulator and asked volunteers from the general public to take part in experiments. Smoke would fill the simulated cabin and help recreate the urgency of leaving a burning plane. Volunteers were provided with an incentive to get out first – £5 was given to the first people out of the plane.

IN CASE OF EMERGENCY

Aircraft interiors and flight procedures are designed differently today thanks to the work of Helen Muir, at Cranfield University. By designing a unique way of evaluating the factors influencing survival in accidents, Muir found ways to increase aircraft safety.

Cabin crews acting more assertively help more passengers

One of the key conclusions of the research is that space is needed next to the airplane exit, and this has been now incorporated in planes across the world. The researchers also showed that cabin crews acting more assertively help more passengers to survive accidents.

The research facility is the only one in the world that is designed to reproduce not only today's but future aircraft conditions. It is used by manufacturers, operators and the organisations that develop safety rules.

COOL RUNNINGS

Every summer commuters on the London Underground face uncomfortably hot temperatures during their journeys, with heat exhaustion on overcrowded trains an increasing problem.

But academics at London South Bank University working with London Underground have developed a revolutionary cooling system that utilises one of the capital's natural resources: rising underground water.

The first prototype of the system is operating at Victoria station. Millions of gallons of water that flow daily through the underground River Tyburn will pass over a cooling unit producing cool air that will be pumped out on to the platform. The water is just a short distance from the Victoria Tube station platforms and promises to reduce summer rush-hour temperatures of 32C by 10 degrees.

The researchers, led by Graeme Maidment, a senior lecturer in engineering at the University, estimate that there is enough water to cool the entire Victoria Line.

The installation at Victoria is likely to be the start of a network wide solution to over-heating on underground trains. The system is ideally suited to the Tube because ground water is most readily available in the deepest parts of the network, which suffer the highest temperatures during the summer months.

Academics developed a cooling system utilising underground water

The water has to be pumped out of the system anyway so it might as well be used to improve the journeys of millions of Tube passengers, argue the academics.

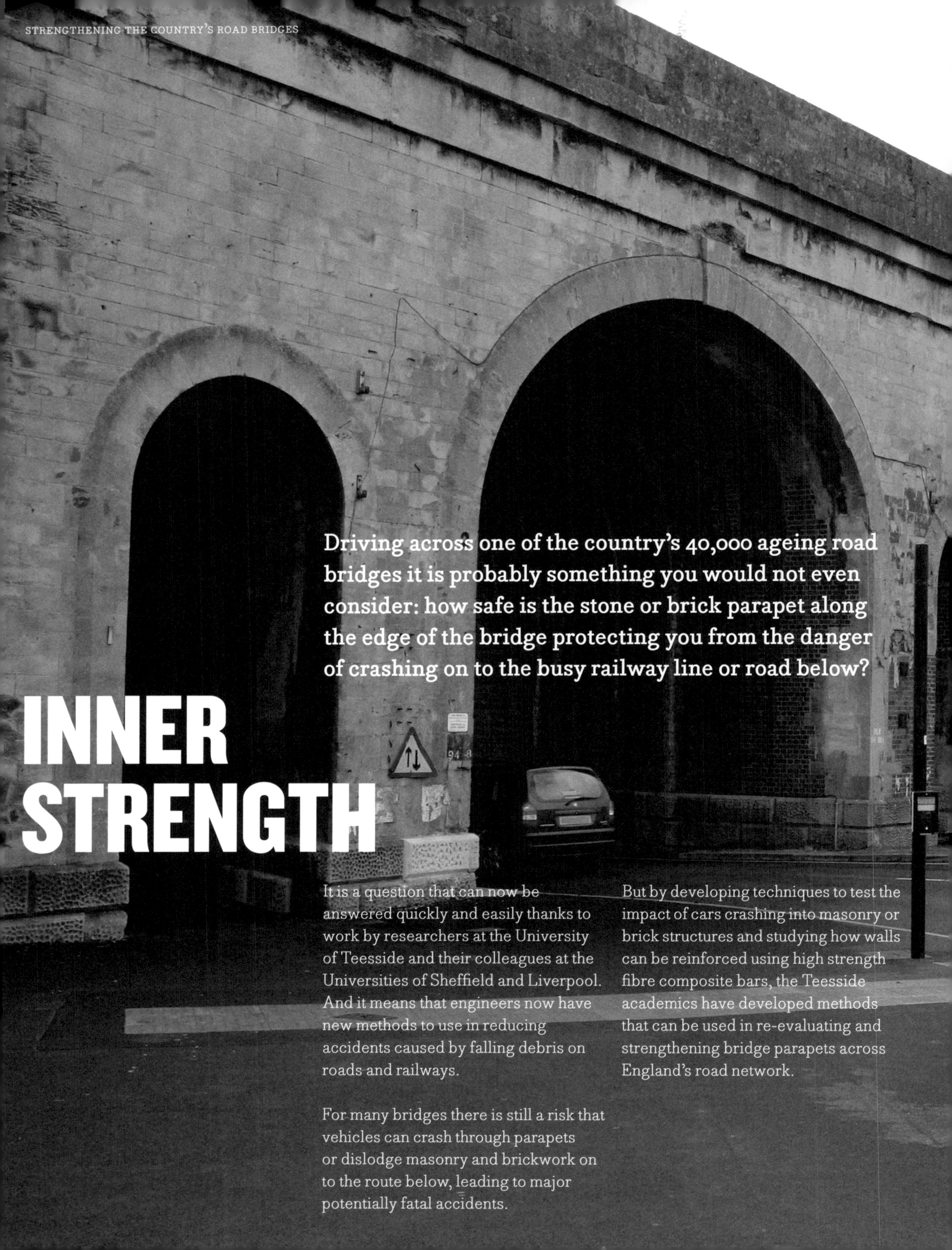

INNER STRENGTH

Driving across one of the country's 40,000 ageing road bridges it is probably something you would not even consider: how safe is the stone or brick parapet along the edge of the bridge protecting you from the danger of crashing on to the busy railway line or road below?

It is a question that can now be answered quickly and easily thanks to work by researchers at the University of Teesside and their colleagues at the Universities of Sheffield and Liverpool. And it means that engineers now have new methods to use in reducing accidents caused by falling debris on roads and railways.

For many bridges there is still a risk that vehicles can crash through parapets or dislodge masonry and brickwork on to the route below, leading to major potentially fatal accidents.

But by developing techniques to test the impact of cars crashing into masonry or brick structures and studying how walls can be reinforced using high strength fibre composite bars, the Teesside academics have developed methods that can be used in re-evaluating and strengthening bridge parapets across England's road network.

Engineers have new methods to reduce accidents from falling debris

With support from the organisations in charge of railway lines such as Network Rail and London Underground, the researchers have developed ways to make more road bridges capable of meeting today's more demanding safety regulations. The aim also is to strengthen the structures, but retain the same external appearance of the bridges often designed more than 100 years ago.

Deborah Withington's research idea came to her in a traffic jam. On hearing a siren, she saw that the drivers around her couldn't tell where the ambulance was coming from.

DIRECTING SIGNALS

The siren can be used to signal emergency exits during fire or power failure

An expert in how the brain responds to sound, Withington identified the problem straight away: the siren was the wrong type of sound.

Conventional sirens are simple sounds covering just one or two frequencies, which makes them difficult to locate. Withington's 'Localizer' siren, developed at the University of Leeds in 1994, uses directional sound.

Our knowledge of what type of sound is accurately located by the human ear has been known for over 100 years. Withington's work was the first time this knowledge was taken from the purely scientific domain and used in applications to benefit us.

It combines a wailing sound to attract attention with a burst of white noise – covering the whole range of frequencies in human hearing – to aid location. Many ambulance, fire and police forces have adopted it across the UK.

The siren can also be used to signal the location of emergency exits, particularly useful during a fire or power failure, when emergency lighting might fail or be ineffective.

Directional sound is now common on reversing vehicles and in 10 years will be widely used in buildings, ships and planes, improving the chances of survival during emergency evacuations.

DEBORAH WITHINGTON
THE INVENTOR OF DIRECTIONAL SOUND

To me it was so obvious. Anyone who has frantically tried to find their ringing mobile phone should understand.

"The whole idea of directional sound is that you can instantly pinpoint it, wherever it is coming from. It first came to me when I was sitting in a car, trying to locate an emergency siren. It was my eureka moment. I had recently read an article about a new driving simulator in the university where responses to a siren could be tested in a controlled environment, so perhaps this was also in the back of my mind. But I do think that certain people have the ability to think laterally.

The science of directional sound – and the fact that sound with multiple frequencies is best located – has been known for 100 years. But no one had thought about how this might be used in the real world. To this day I cannot understand why no one else had thought of it before. To me it was so obvious. Anyone who has frantically tried to find their ringing mobile phone should understand.

I was told that I wouldn't get a research council grant because I was dealing with a potential product, not basic science. Instead I had to set up a company. Things take an order of magnitude longer than you think they will in business. Companies in the UK and the US have taken up licenses, although it took three years to finalise the US deal.

My passion is applying the idea to evacuation. You are in a burning building, where do you get out? Often, you don't. Directional sound has massive life-saving potential.

We are currently developing a new cane for blind people. The idea came out of a group discussion on ultrasound, which brought together a food scientist, an electrical engineer, a biologist who studied bats, and a neurophysiologist. The cane works by sending out ultrasound, which bounces back and through vibrations, can tell the holder which obstacles lie ahead.

How would I help other projects like mine? Set aside a percentage of the science budget to create a stream for promising areas of research. Everyone can argue about the percentage; but the point is that you would have two streams for basic and applied research running at the same time. Get a few people with an entrepreneurial track record, people who have actually done it. Pay them properly. And ask them to assess whether to take ideas forward.

The stream should not be based in the research councils or the Department of Trade and Industry, but in some separate intermediate body. Some ideas will not work; some will come through. These should be captured quickly, and nurtured properly."

IDEAS FOR IDEALS

We all dream of a fairer world, where poverty and famine do not exist, where schools provide a decent education for every child, where justice is balanced, where politics empowers people irrespective of where they come from, and where support is available for the elderly. But it is UK academics who have shown that powerful ideas are needed to overturn previous myths and dogmas, and help to improve people's lives.

Social scientists have improved our understanding of the nature and impact of poverty, transforming the way help is provided for the less well-off. For example, poverty, not family circumstance, is a cause of learning difficulties among children from one-parent families.

Economists in UK universities have focused their minds on solving the problems in the developing world. We now understand far more the reasons that lead to poverty in poor countries, while national accounts help to sustain developing economies. Amartya Sen, meanwhile, transformed our view of famines, showing that they are due not to lack of food, but lack of access to food. And thanks to another academic we are beginning to tackle modern day slavery.

Academics also uncovered a crucially important finding that has transformed Government education policies: schools in poor areas can make a difference in raising educational standards (although only so much can be done with intakes of poor students to start with). On the other side of the age spectrum, wage-related state pensions were inspired by the ideas from UK academics.

One of the leading legal philosophers of the 20th century meanwhile wrote the definitive text reminding lawyers of the difference between morals and laws.

Researchers also produced the groundwork for the creation of the new Parliament in Scotland and the National Assembly in Wales, and charted the changing nature of voting trends across the UK. The 'Third Way' meanwhile offered a new approach to politics altogether, arguing that neither traditional Left and Right wing politics can offer a way forward in the modern, globalised economy.

The findings have transformed the way that Government and charities provide help for the less well-off.

The link between deprivation and ill-health has been known for many years, prompting the Victorians to introduce clean water systems and sanitation. But thanks to the work of Thomas McKeown at the University of Birmingham in the 1970s we are now much more aware of all the influences of social conditions on health.

McKeown looked at the number of deaths in England and Wales caused by different diseases from the 19th century until the early 1970s. His statistics revealed a huge impact of factors such as nutrition, water supply, sanitation, food hygiene, smoking, diet and exercise on the health of less well-off communities. The study has led the Government to take far more seriously these factors in its efforts to improve the health of people.

During the 1970s Richard Morris Titmuss at the London School of Economics and Political Science meanwhile found that poverty, not family circumstances, were behind the behavioural problems and learning difficulties in children from one-parent families.

In the late 1970s and early 1980s Peter Townsend at the University of Essex, revealed the full reality of social inequalities among different communities. Townsend identified the key groups living in poverty, including: the unemployed, low paid workers, disabled people and the long-term sick, large families, one-parent families, families with children with disabilities, older workers and the elderly. He also showed that the deprived do not only have poorer housing and diets than the better off, but also face more limited lives at work and within the family.

IMPROVING LIVES

What does it mean to be poor? Social scientists in UK universities have produced a series of seminal studies that have improved our understanding of the nature and impact of poverty.

In the 1990s Michael Noble at the University of Oxford produced new official statistics that can more accurately identify those living in disadvantaged communities. The National Index of Multiple Deprivation covers several 'domains' of deprivation: income, employment, health, education, housing and access to services. The Index is used to allocate more than £2 billion of Government spending every year in the UK.

The study has led Government in its efforts to improve the health of people

His outstanding work was the *Theory of Economic Growth*, published in 1955 when he was at Manchester. It was an exhaustive analysis of the economic, political and cultural factors that contributed to the development of poor countries.

Lewis tackled issues central to causes of poverty in the developing world

Lewis tackled issues central to the causes of poverty among populations in the developing world and to the unsatisfactory rate of economic development. His two renowned theoretical explanatory models, designed to describe and explain the intrinsic problems of under-development, formed the basis of debates that continue today.

He was also interested in the problems that arise when central planning ignores 'price signals' from market systems. He stressed the distinction between 'planning by direction' and 'planning through the market'.

GLOBAL VS LOCAL

Arthur Lewis, at the London School of Economics and Political Science and at the University of Manchester, led economists to study the reasons for the poverty and development problems of poor countries. His research showed how relations between local agriculture and modern markets combined to create poverty.

A doctor assessing the health of a patient can do many things: check his pulse, his temperature and ask him how he feels. But given the billions of transactions that take place in the world today, how can economists assess the economic health of a nation?

HEALTHY ECONOMICS

The answer was provided by Richard Stone during the 1950s at the University of Cambridge, when he created the methods needed to produce national accounts. His approach has been taken up internationally (but not in the UK), helping to create the economic policies that sustain most developed and developing economies.

Stone's innovation provides a better basis for economic analysis and policy

Stone's innovation was to produce a logically connected system of national accounts, where each sector is dependent on the other. This provides a much better basis for economic analysis, forecasts and economic policy. His accounting methods have been especially helpful to international organisations such as the United Nations and the World Bank as they try to create the conditions of economic growth around the world.

STARVATION NATIONS

'Starvation is the characteristic of some people not having enough food to eat. It is not the characteristic of there being not enough food to eat.' With this simple observation, the economist, Amartya Sen, changed our view about famines.

Many people still believe that famines are caused by a shortage of food. But in the 1980s, while at the University of Oxford, Amartya Sen showed this was not in fact the case. He was the first to conclude that people suffer in famines not because of food shortages, but because they lack the resources or other entitlements that are needed to obtain food.

Sen noted how, even in poor and hungry regions, there is often sufficient food, shown, for example, by the export of agricultural produce from afflicted regions.

For economists and policy makers, this was a new insight, which explained some of the terrible famines that have occurred in the past. It also pointed towards a way to create a world without famine – for example, to provide entitlements to those who lack them at the time, and create a worldwide insurance policy to compensate those in need of aid.

They lack the resources or other entitlements that are needed to obtain food.

There are 27 million slaves in the world today – more than at any time during the 5,000 years that slavery has existed. But only now are we beginning to tackle modern day slavery thanks in no small part to the work of a sociology professor at Roehampton University.

FREEDOM ROAD

Professor Kevin Bales' 1999 book *Disposable People: New Slavery in the Global Economy* has brought about a new awareness, new laws and new programmes for the liberation and rehabilitation of slaves around the world.

Bales' arguments have helped to inspire new policies on slavery

Bales' work has been instrumental in highlighting the plight of children forced to work in industries in south Asia and north Africa – often exploited by companies in the West. The key to combating slavery is to recognise it as a global economic problem, and that both slaves and slave owners need rehabilitation, support and education.

Bales' arguments have helped to inspire new policies on slavery and human trafficking for the Nepali, Norwegian, Irish, British and US Governments, west African States and within the United Nations. Laws in Nepal led to the release of 40,000 bonded labourers.

TAILORED CLASSES

In the 1960s and 1970s, Basil Bernstein, at the Institute of Education, University of London, showed how the design, organisation and control of school lessons should be tailored to suit particular children.

Teachers became more aware of how to get the best results from children

Bernstein found that children from diverse social backgrounds responded differently to what they heard in school. As a result, teachers became much more aware of how to alter the content, pace and order of lesson materials and discussions to get the best results with children from different backgrounds.

According to Bernstein there are two distinct codes of language – 'restricted' and 'elaborated' forms. Families from poor backgrounds tended to talk in the former code, using short sentences and a limited range of topics. Middle class families used more complex sentences in a code with a wider range of uses and which can encompass more abstract ideas. This, Bernstein claimed, explained the relative under-achievement of working class children in school.

Bernstein's work also challenged the practice of 'streaming' in schools where children are grouped in terms of ability.

In work begun in 1967, and continued at the University of Oxford from 1968, James Mirrlees found methods for analysing how tax systems can be used to give incentives to people even though the desired behaviour can't be observed.

PROGRESSIVE TAXMAN

A solution for introducing progressive taxes

One example, published in the 1970s, was a theory of optimal income taxation. It provided a solution for introducing progressive taxes without having incentives to work. Another was the calculation of optimal insurance contracts when there is 'moral hazard' – a lack of care because consequences are uncertain.

These methods in the 'economics of asymmetric information' have been used by other economists in a variety of applications.

SUCCESSFUL POOR

It was not so long ago that people assumed that schools in poor areas would always under-perform, and enrolling working class children would lead automatically to poor academic results.

But in 1979 Michael Rutter and fellow researchers at the Institute of Education, University of London shattered this myth; they showed that schools in poverty-ridden areas could be successful and revealed the secrets of their success.

They found that children prospered in schools where there was an emphasis on an academic culture, clear expectations and regulations, high levels of student participation, library facilities, vocational work opportunities, and extracurricular activities.

The success factors identified in the research have been used by teachers

The longer students attended these successful schools, the less severe their problem behaviours became. In contrast, for unsuccessful schools, the opposite was true – the longer students attended them, the more unruly, disruptive and delinquent they became.

The success factors identified in the research have been used by teachers to create better learning environments for children all over the UK and beyond, and impacted on Government education policies. Although the researchers stressed that academic results are still largely determined by the type of children schools enrol to begin with.

Pensions have become headline news in recent times, as people have found that 'guaranteed' funds for retirement have been lost by companies in the financial markets. And as the number of people living longer continues to rise, Governments are faced with major dilemmas.

STATELY PENSIONS

That there are wage-related state pensions at all is down to UK academics

For the old in the 1950s, there were only very basic pensions. Then Brian Abel-Smith and Peter Townsend from the London School of Economics and Political Science made the case for wage-related state pensions, changing the course of legislation in the UK such as the Pensions Act of 1959.

Their work also influenced pension, social security, unemployment and sickness-related benefit schemes in the 1960s and 1970s.

LEGAL POSITIVISM

The work of HLA Hart (Herbert Lionel Adolphus Hart), one of the most important legal philosophers of the 20th century, changed the way that lawyers understand their world and their work. Hart argued that law and morality are independent but interconnected.

He has helped generations of lawyers to steer a path between the twin vices of legal cynicism and legal idealism – how to distinguish legal rules from moral ones without reducing the legal rules to exercises of raw power. Hart, who was Professor of Jurisprudence at the University of Oxford, revolutionised the methods of jurisprudence and the philosophy of law in the English-speaking world.

Hart's most famous work *The Concept of Law*, first published in 1961, developed a sophisticated view of 'legal positivism'. It is seen as one of the most important books written in the field of legal philosophy. Each chapter has been the springboard for entire areas of discussion since its publication, such as law as a system of rules, and the separation of law and morality. This has helped lawyers be more self-aware in their work.

One of the public legacies of Hart's work is that the Supreme Court of the United States, and more recently the Judicial Committee of the House of Lords, have come to treat the nature of their own activities as a matter of open debate in their judgements.

Hart revolutionised the methods of jurisprudence and the philosophy of law

Today Scotland and Wales have their own national political institutions within the UK. These constitutional reforms were greatly facilitated by the groundwork of Professor Robert Hazell and his team of researchers in the Constitution Unit at University College London.

CONSTITUTION SOLUTIONS

The contribution of the Constitution Unit has been quite remarkable

The researchers helped to lay the foundations for devolution in Scotland and Wales, the Human Rights Act, reform of the House of Lords, freedom of information, and changes to the voting system. Taken together these changes amount to a new constitutional settlement affecting the daily life of every person in the UK.

In a series of major reports to ministers and civil servants the Constitution Unit focused on the design, implementation and consequences of constitutional reform.

The Unit's work is frequently referred to in political and parliamentary debates. Robin Cook MP, when Leader of the House of Commons, said in 2001 "It would have been much more difficult to have moved so rapidly and extensively if we had not benefited from the expert and detailed work on the reform programme by the Constitution Unit". And more recently Lord Falconer, Secretary of State for Constitutional Affairs, said "The contribution of the Constitution Unit, and its director, Robert Hazell, on these issues has been quite remarkable".

No UK election news broadcast is now complete without the 'Swingometer'. But few people realise the idea of election swing is an academic concept – created by David Butler of the University of Oxford and other researchers.

SWINGOMETER

Using the notion of swing it is much easier to understand why elections turn out the way they do.

Researchers have also charted the changing nature of voting trends. The longest running series of academic surveys in the country is the British Election Study (BES), currently based at the University of Essex, which has surveyed a representative cross section of the electorate at every general election (and referendum) since 1964.

It has traced the gradual weakening of party loyalties, widening regional differences, and the recent drop in turn-out of voters for elections. One finding is that it is much harder to predict how people from different social back-grounds will vote in modern elections.

Most of the surveys re-interview respondents from previous surveys, allowing the researchers to provide an accurate analysis of shifts in voting patterns and political opinions. It has also studied the impact on voting of long-term changes in society, manifestos, styles of campaigning, the role of party leaders, and the influence of the media.

It is much easier to understand why elections turn out the way they do

The approach was promoted by Anthony Giddens, former director of the London School of Economics and Political Science, who argued that new political solutions were needed to respond to the modern globalised and constantly changing world we now inhabit. Giddens' ideas influenced the policies of the UK Government, under Tony Blair, and the US administration, under Bill Clinton, and a large number of other centre-left Governments across the world.

ALTERNATIVE APPROACH

The 'Third Way' – an alternative approach to traditional Left or Right Wing politics – is one of the main ideas that inspired the centre-left policies of Governments during the late 1990s.

Giddens argued that we must rethink how the Welfare State operates

According to Giddens, conventional politics cannot address the inequalities created as a result of three inter-connected changes in modern society: globalisation; 'de-traditionalisation' (the need for traditional institutions, such as the family, to justify themselves); and 'social reflexivity' (the replacement of past certainties by the need for people to take more decisions for themselves).

The 'Third Way' accepts the importance of market competition and the need to create a dynamic economy but combines this with emphasis on the importance of social justice. Giddens argued that we must rethink how the Welfare State operates, shifting from the traditional passive and reactive approach to an active one where Government provides security to allow people to take risks.

MICHAEL RUTTER THE CHILD PSYCHIATRIST WHO FOUND THE REASONS FOR SCHOOL EFFECTIVENESS

The response from schools [to the study] was very positive, but academic educationalists were outraged

"The story began when we noted a marked variation among [the performance of] schools in a previous Isle of Wight study. That epidemiological [medical] study was not focused on schools at all and the observation was an incidental one. Nevertheless, it was obvious that it was of potential importance and that the key question was whether the variation reflected differential intakes to the schools or differences in school effectiveness.

The Isle of Wight authorities were not at all positive about the study. To begin with, they were sceptical as to whether schools would co-operate and, secondly, they considered that you did not need a study in order to know what was good or bad about a school. But then exactly the same issue came up in our London study. Another spur was the forthright questioning by a prominent Head Teacher in the area.

The research gained its strength from an active collaboration among disciplines. The fact that the planning for the research involved teachers as much as researchers meant that we did receive the co-operation of schools. When a teacher's strike intervened in the middle of the study, the teachers were instrumental in ensuring that the strike did not interfere with the research. There is no doubt that that made all the difference.

Then there was the collaboration among different disciplines within our research group. The team was made up of a psychologist who had been a practising teacher, a developmental psychologist, and someone whose background was in social work and social administration. I was primarily an epidemiologist and psychiatrist. I think it was a real plus that I came from outside because it enabled me to bring knowledge of research in other arenas and because it enabled me to pose awkward questions.

The main funding came from the [then] Department of Education and Science. Obviously, the DES will have sought advice but, so far as I know, the application never went out for peer review as such. If it had, it would pretty certainly have been rejected out of hand. It would never get funded today by a research council.

There is no way in which educationalists would have supported a study on schools that was undertaken by a mere child psychiatrist. The response from schools [to the study] was very positive, and so was the response from the broader psychological community, but academic educationalists were outraged.

The Government (both at that time and now) were very willing to accept the message that even the schools in socially disadvantaged areas could make a difference, but they were totally opposed to all the findings that had to be taken on board with it. Thus, the balance of intake to schools did make a difference and no Government has been willing to accept the implications of that. Also, there was reluctance to appreciate the realities of the time it takes to bring about change.

I do see a place for strategic decision making in research by funding bodies but, equally, I would see the first priority (always) as being recognition of the creative, innovative ideas of top class individual researchers. Virtually all important scientific advances derive from such individual initiatives and not from committee decision making."

UNDERSTANDING OURSELVES

Academics in the arts and humanities are sometimes thought of as the 'antennae' of society, revealing our changing values, attitudes and beliefs. The last 50 years have seen the certainties of previous ages – religious and political beliefs, national identities – disappear for many people, making our sense of identity perhaps even more important.

History books and programmes have always been popular despite the sometimes uncomfortable messages they tell: all Governments should share the blame for the Second World War (AJP Taylor); peace is a relatively modern idea (Michael Howard); and capitalism faces an uncertain future (Hobsbawm).

For all the changes in society, an architect deciphering the first European language revealed that people 4,000 years ago had much in common with the modern world. According to the scripts on ancient stone tablets found in modern-day Crete, for example, the early Europeans operated a complex system of land ownership. Other studies meanwhile have shattered our preconceptions of our European forebears two and a half thousand years ago – now known as the Celts.

Yet, according to the philosopher Karl Popper, no matter how much we know about the past, history cannot be used to predict the future. The advice was picked up eagerly by Right Wing politicians to support free market policies of the 1980s.

Literary studies meanwhile have revealed how society impacts on language and story-telling. Researchers have also played a key role in documenting our heritage for future generations, producing for example the first definitive guide to the vocabulary of the Welsh language.

Pevsner's architectural guides produced the first authoritative source of information on the country's great buildings, dubbed the 'the greatest endeavour of popular architectural scholarship in the world'.

COSTLY BLUNDERS

"Human blunders shape history more than human wickedness," Alan John Percivale (AJP) Taylor said. His seminal work, *Origins of the Second World War* (published in 1961), changed our perceptions of the war forever.

Taylor argued the war resulted from indecision by Britain and France

Taylor argued that Adolf Hitler and the rise of Nazism should not be solely blamed for the conflict. It was triggered not because of the Third Reich's intentions, or a clash between good and evil, but the indecisive policies and actions of Great Britain and France towards Hitler.

Taylor, a Fellow of Magdalen College, the University of Oxford, was the harbinger of uncomfortable truths for all Governments involved in the war.

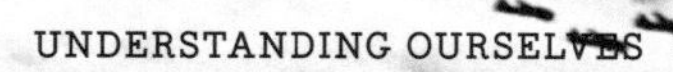

Allowed to study official files from the Reichstag in Berlin, he produced an exhaustive analysis of British and German diplomacy leading up to the war, much of which contradicted the official public version of events.

Taylor argued that the war was not unique, and was not caused by the rival ideologies of Fascism, Communism and Liberalism, or Hitler's dream of world domination. It happened because of blunders on the part of Britain and France and opportunism on the part of Germany.

Hitler's anti-Semitism was not unique; the German leader took advantage of the mood throughout Europe. Britain pursued its own national interests, while Poland was weak, corrupt and elitist, argued Taylor. The United States meanwhile was almost totally isolationist in the run-up to the war.

According to Taylor, Hitler was a 'traditional European statesman' seeking to restore Germany. He 'simply leaned on the door hoping to gain entrance and the whole house fell in'.

analyse the factors behind the latest wars around the world, asking whether peace will ever be possible.

Only after two world wars did peace become an objective of countries

Howard's books remind us that it was not until the Enlightenment of the 18th century that war came to be regarded an unmitigated evil. Before this, societies took war for granted.

It was only after the massive slaughter of two world wars during the 20th century that peace became the declaratory objective of 'civilised' countries.

Nevertheless war in one form or another continues in the modern world, prompting Howard to pose many unresolved questions. Is war in some way still a necessary element in maintaining international order? Are war and peace dependent on each other?

Howard's expertise in warfare has been gained through practical experience as well as study. He served in Winston Churchill's Personal Security Detail before earning a Military Cross at Salerno during World War Two.

"Peace is a modern invention; war is as old as mankind." These words sum up one of the major dilemmas for society raised by the work of Michael Howard, who 40 years ago became the country's first Professor of War Studies at King's College London.

A CHANCE FOR PEACE?

The sheer breadth and popularity of Eric Hobsbawm's work marks him out as one of the world's great historians. Hobsbawm, now Emeritus Professor at Birkbeck College, University of London, has charted the complex patterns and mechanisms that transformed the world during the 19th and 20th centuries.

WATCHING THE WORLD

Hobsbawm's commentaries cover 240 years of modern history, spanning the Industrial Revolution, the rise of the British Empire, the world wars and the Cold War, the rise and fall of Communism and the inexorable growth of capitalism in the world over the period.

His trilogy of books – *The Age of Revolution* (published in 1962), *The Age of Capital* (1975) and *The Age of Empire* (1987) – describe the 'long 19th century', from 1789 to 1914, and is seen as the defining work for this period of history. *The Age of Extremes* (1994) meanwhile covers world events during the 'short 20th century' from the First World War to the collapse of the Soviet Union in 1991.

His trilogy of books is seen as the defining work

In 1953 a young British architect and a University of Cambridge scholar showed how they had unlocked Europe's earliest known language, unread for 3,000 years.

UNLOCKING SECRETS

The inscriptions on the tablets suggest that the ancient Minoan civilisation on Crete had been in contact with the Greeks far earlier than believed. The speakers of the first European language lived in a highly organised society, with kings, warrior castes, slaves, priests, local council officials, and a complex system of land ownership.

Speakers of the first European language had a highly organised society

Michael Ventris and John Chadwick had conquered what came to be known as 'the Everest of Greek archaeology'. The breaking of the ancient code called 'Linear B' shattered many of our preconceptions about early Western civilisation.

Ventris was a 30 year-old amateur enthusiast in the art of ancient code breaking, but he managed to unravel a puzzle that had perplexed academics for decades: the secret key to scripts etched on clay tablets discovered in Knossos, in modern Crete. Chadwick, a University of Cambridge expert in the history of the Greek language, played a crucial role in confirming Ventris' discovery.

Linear B dates back to the century between 1300 – 1200 BC, and was used long before Homer's accounts of the Trojan War were written down.

Ventris and Chadwick published their joint work, *Documents in Mycenaean Greek*, in 1956 just after the tragic death of Ventris in a car accident at the age of 34. The work is still published by the Cambridge University Press, and the University of Cambridge retains the most important Linear B reference archives in the world.

SOPHISTICATED CELTS

Ever since Roman times, the people of non-Mediterranean Europe, whom some ancient writers called Celts, have been stereotyped as primitive and uncivilised war-mongering tribes.

In fact our European forbears two and a half thousand years ago were in many ways as sophisticated and intellectual as the civilisations of Rome, Egypt and the East.

There is evidence of such sharing of ideas, particularly in art and religious beliefs, that these peoples may have seen themselves as having a European identity: the Celts embraced a complex European-wide belief system that included perceptions of spirituality throughout the natural world, including the sun, moon and water. They also held that the world was sacred and needed to be respected and protected (not so very different from Green concerns today).

These are the messages emerging from the work of Professor Miranda Aldhouse-Green, Professor of Archaeology at University of Wales, Newport.

The Celts embraced a complex belief system that included spirituality

Through books such as the *Dictionary of Celtic Myth and Legend*, *The World of the Druids*, *Dying for the Gods* and lectures across the world, Aldhouse-Green has helped to transform our perceptions of the ancient Celts.

Combining archaeological research and the testimonies of classical writers, the studies reveal a subtle, deeply spiritual and sophisticated people, who produced breathtakingly beautiful and complex art and possessed a highly-developed religious system presided over by a powerful, pan-European priesthood, known as the Druids. These religious leaders were not only concerned with sacred matters but were also politically influential and skilled as healers, judges, poets and scientists.

THE OPEN SOCIETY

One of the great philosophers of the 20th century, Karl Popper helped to shape British politics in the 1980s and also changed our views of how science develops.

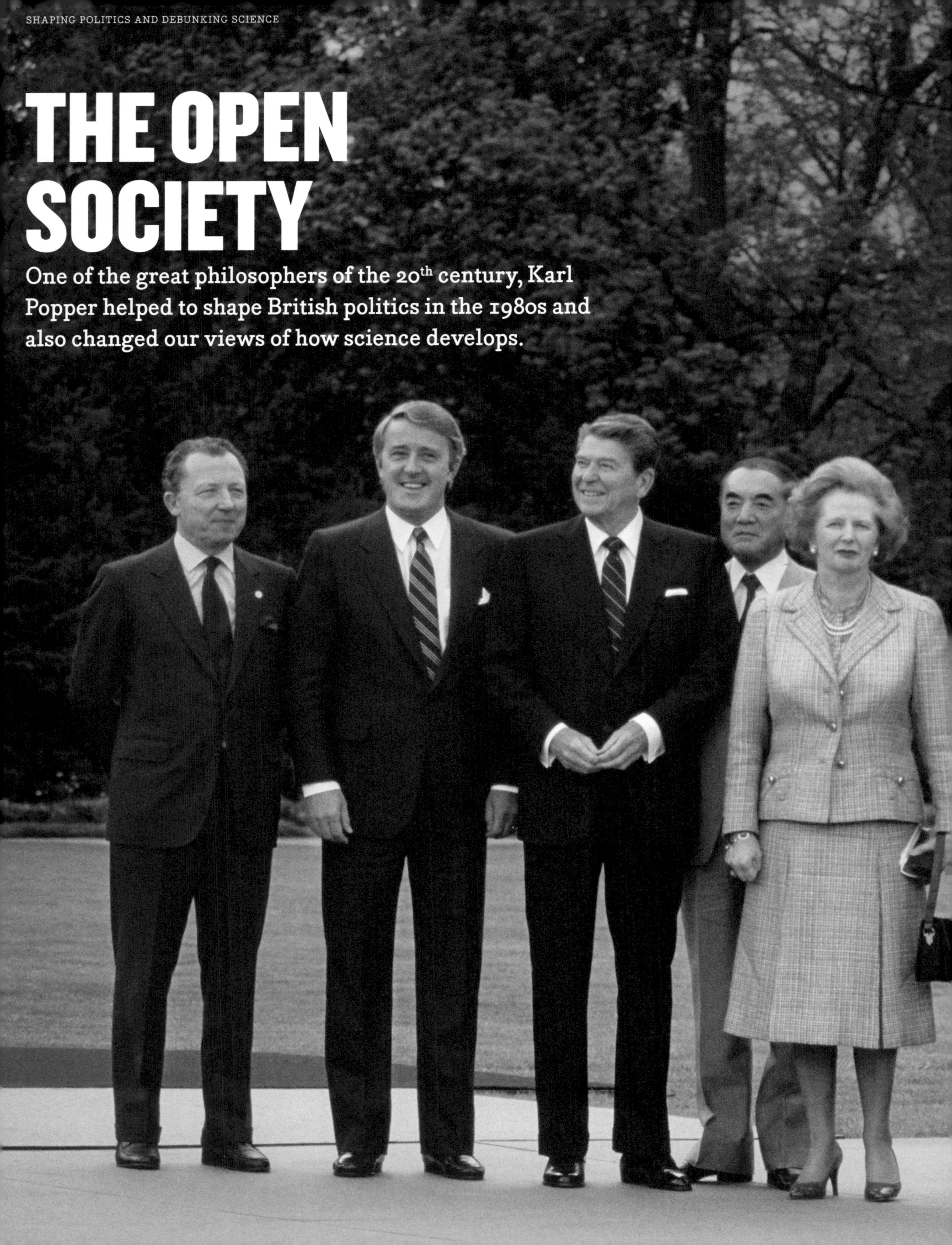

History cannot be used to predict the future, argued Popper, who was Professor of Logic and Scientific Method at the London School of Economics and Political Science.

His attacks on attempts to plan society, witnessed during the century in former Communist countries, inspired many of the free market policies adopted by leaders in the West.

Popper's ideas as well as those of the economists Friedrich von Hayek and Milton Friedman hugely influenced the policies of the 1980s Conservative Government led by Margaret Thatcher.

Popper postulated that science progresses by daring theoretical leaps

Popper's anti-Marxist book *The Open Society and Its Enemies* has been called one of the most influential books of the century. He questioned the existence of inexorable laws of human history, arguing that history is influenced by the growth of knowledge, which is unpredictable.

But much of Popper's work concerned the way that science works. In his book *Conjectures and Refutations* he postulated that science progresses by daring theoretical leaps, not through the steady methodical building up of data.

Popper pointed out that scientific theories cannot be completely verified – only refuted when new evidence emerges that contradicts former predictions.

POPULAR CULTURE

Written and spoken commentary on books, plays and TV programmes by UK academics during the last 50 years reveal much about changing values and attitudes over the last half century.

Richard Hoggart's 1957 work of literary sociology, *The Uses of Literacy* stands as a pioneering study of what had been achieved, since the Universal Education Act of 1870, and the Butler Education Act of 1944, for the working classes of Britain.

Had their culture – the values of the working class community (where Hoggart himself originated) – been enriched by the greater access to reading materials and greater proficiency in reading them? Hoggart, who was later Professor of Modern Literature at the University of Birmingham, asked what reading was actually for. His best-selling book triggered a debate on the issue of 'culture', and remains a widely read and still influential text.

Frank Kermode's 1967 book *The Sense of an Ending*, jolted British literary criticism into unknown territory – the dangerous realm of 'theory' as it was called. The book begins with what seems like a simple question: why do we like our narratives to 'finish' – to have literally, as with old movies, the statement 'the end' appended to them?

Famously, Kermode, a professor at University College London and the University of Cambridge, pointed out that when a clock goes 'tick tick', we hear 'tick tock'. He argued that this is because we are wired to frame everything as beginnings and endings.

Meanwhile in a series of books beginning with *Culture and Society* (published in 1957) Raymond Williams established himself as one of the leading critics in the UK during the late 20th century. His aim was to re-insert literature back into the history and society from where it originated.

He showed how new attitudes and values of society change our interpretation of words in the English language. The word 'popular' for example was first understood during the 15th century as a negative term, meaning 'low' or 'of the common people'. By the turn of the 20th century however the word had taken on its more modern positive meaning – 'liked by all'.

The change reflected the move away from an elitist society to one that is now more inclusive and egalitarian. The Professor of Drama at the University of Cambridge went on to show that plays and TV programmes similarly mirror developments in society.

New attitudes and values of society change our interpretation of words

On 6 December 2001 the final draft entry was written after 80 years of labour, over half a century of drafting entries, and the completion of two million citation slips. So ended the most important project associated with the Welsh language in 50 years.

DEFINING WELSH

The first standard historical Welsh dictionary, *Geiriadur Prifysgol Cymru*, was complete.

The dictionary will have a similar impact on the Welsh language community to that of The Oxford English Dictionary on the English-speaking world. It is the ultimate authority on the spelling, etymology, and vocabulary of Welsh.

It is the authority on the spelling, etymology, and vocabulary of Welsh

Edited by academics, led by the late RJ Thomas, Gareth Bevan and Patrick Donovan at the University of Wales with help from hundreds of volunteers, the dictionary presents in alphabetical order the vocabulary of the Welsh language from the remnants of Old Welsh, to the more recent Medieval and Modern periods.

The vocabulary is defined in Welsh, but English equivalent words are also given. All future Welsh dictionaries will use it as the standard reference text.

The dictionary will also revolutionise Welsh literary studies, enabling developments in new fields of linguistic studies, such as Welsh word formation, and Welsh terminology and translation.

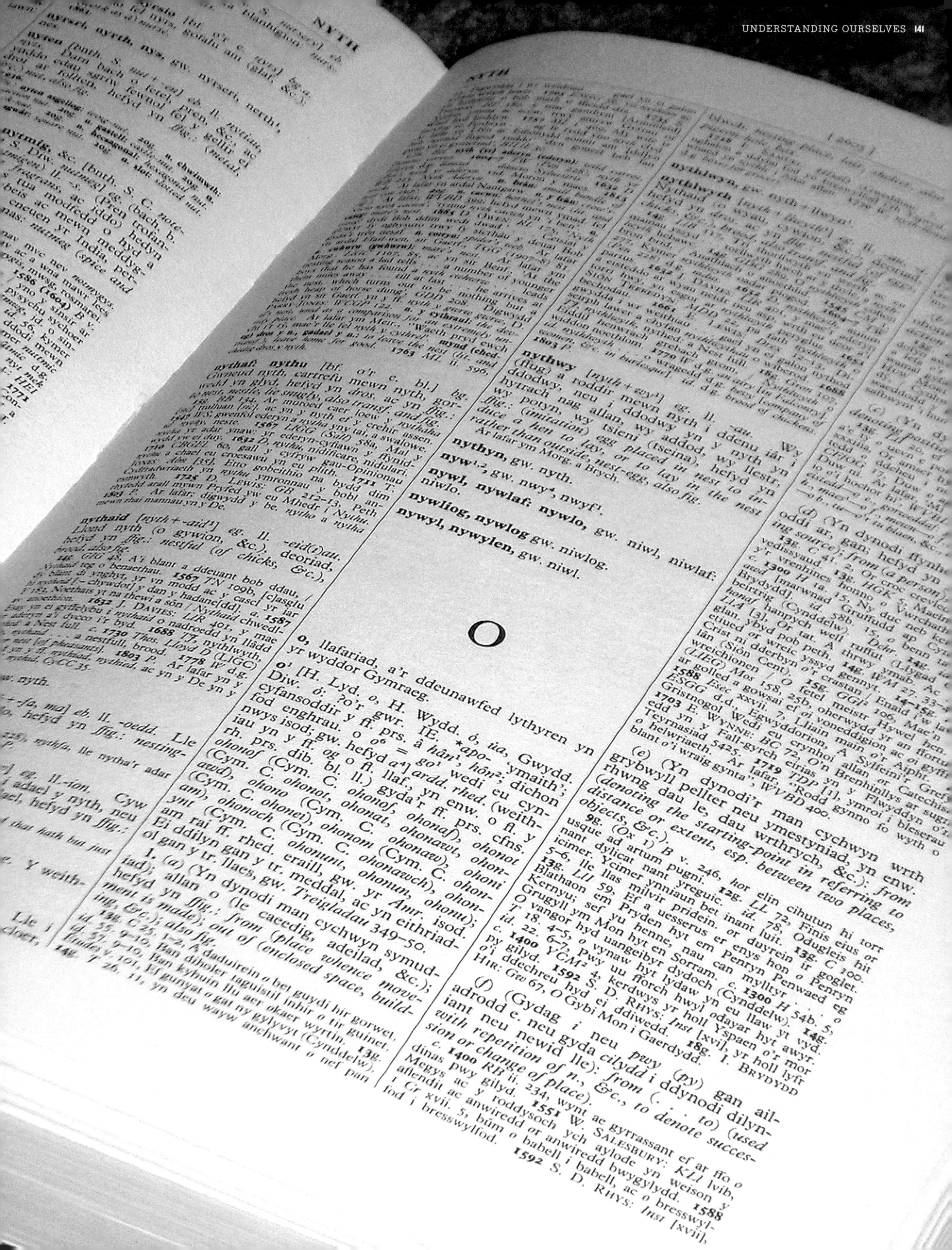
NYTH
nythwy [nyth + gwy¹] eg. ll. -au. Wy (ffug) a roddir mewn nyth i ddenu iâr i ddodwy, neu i ddodwy yn y nyth yn hytrach nag allan, wy addod, wy llestr, wy potyn, wy tsieni (tseina), hefyd yn ffig.: (imitation) egg placed in nest to induce a hen to lay, or to lay in the nest rather than outside, nest-egg, also fig.
Ar lafar ym Morg. a Brych.
nythyn, gw. nyth.
nyw¹·², gw. nwy⁴, nwyf¹.
nywl, nywlaf: nywlo, gw. niwl, niwlaf: niwlo.
nywliog, nywlog gw. niwlog.
nywyl, nywylen, gw. niwl.
O
o, llafariad, a'r ddeunawfed lythyren yn yr wyddor Gymraeg.

ARCHITECTURAL APPRECIATION

Dubbed 'the greatest endeavour of popular architectural scholarship in the world', the Pevsner Architectural Guides have opened the eyes of generations to the design and history of great buildings in Britain and Ireland.

Created by the architectural historian Nikolaus Pevsner, the guides were the first authoritative source of information on the architectural sites that populate the country, from ancient cathedrals, great country houses and their parks to Victorian public buildings and industrial monuments. The guides have helped to raise awareness of the country's rich and unique architectural heritage.

Pevsner, who was based at Birkbeck College, University of London, took 25 years to complete the volumes, which cover all the English counties. The first county he toured was Middlesex. The first book, on Cornwall, appeared in 1951. And the 46th and last, covering Staffordshire, was published in 1974. Books covering Wales, Scotland and Ireland also followed.

Each Pevsner guide includes an introductory overview of the distinctive architecture of the area, and an alphabetical list of places of architectural interest. New editions of the guides continue to be published today, keeping the series up-to-date with new information on older buildings and recent architecture.

The guides raised awareness of the country's architectural heritage

MICHAEL VENTRIS
THE MAN WHO DECIPHERED EUROPE'S OLDEST LANGUAGE

The defining moment for Michael Ventris came when as a 14 year-old school boy he visited the 1936 exhibition celebrating the 50th anniversary of the British School at Athens. It was here that the famous archaeologist Arthur Evans revealed the ancient stone tablets he had discovered in Crete two decades previously.

Etched on the tablets were the 3,500 year-old symbols of the first language in Europe, named Linear B. A teacher remembers Ventris asking: 'Did you say the tablets haven't been deciphered, Sir?' Thus began a life's obsession.

Ventris showed an early talent for languages, learning several by the age of 10. He was also fascinated by ancient scripts, buying a book on Egyptian hieroglyphs when he was seven. Four years after his trip to Athens, he published his first article on the topic.

He trained as an architect (designing a family house in Highgate, London) and served in the Royal Air Force as a navigator during the war. But the obsession with one of the great puzzles of archaeology and linguistics of the time would not go away.

In 1950 Ventris circulated a Mid-Century Report on Linear B, and then gave up his architectural job to work on the problem of deciphering the ancient code. By May 1952, he reported that the code was 'breaking'. To his astonishment, the Linear B documents were written in an early form of Greek. Ventris announced his discovery to the world on BBC radio in July 1952.

The broadcast attracted the interest of John Chadwick, a lecturer in classics at the University of Cambridge. The two scholars collaborated over the next four years to produce the publication describing the translation in full. But weeks before *Documents in Mycenaean Greek* was published Ventris tragically died, driving his car into the back of a lorry. He was 34. But during his last year he had complained of 'losing the meaning of his life'.

Ventris discovered the Linear B documents were written in an early form of Greek

UNDERSTANDING OUR ENVIRONMENT

The concept of the Earth as a self-regulating, living organism – the 'Gaia hypothesis' – is said by some to be the most powerful idea since Darwin's Theory of Evolution. James Lovelock's ideas have had a huge influence on attitudes towards the environment, now a source of major public concern.

Lovelock is one of several UK academics who, over the last 50 years, have improved our understanding of the Earth's changing atmosphere, climate and oceans.

Lovelock invented the electron capture detector, an incredibly sensitive probe that showed for the first time the extent of chlorofluorocarbons (CFCs) in the atmosphere.

The findings helped to raise public consciousness of environmental issues. But it was the late Hubert Lamb who helped to establish climate change and global warming as serious subjects for discussion. Lamb was one of the first experts to argue that the Earth's climate was not in fact stable on timescales relevant to modern humankind, as had widely been assumed before. Other academics have shown the long-term impact on people of flooding – an increasingly common occurrence both home and abroad.

In 1963, two British marine geologists identified sea floor movement, a discovery that helped to confirm the now accepted theories of continental drift and plate tectonics, describing how the world's continents float on a sea of lava beneath the Earth's surface.

It was while studying the atmosphere of the planet Mars that James Lovelock developed a revolutionary way of thinking about the Earth.

MOTHER EARTH

The 'Gaia hypothesis', the idea of the Earth as a self-regulating living organism, transformed public attitudes towards the environment.

Lovelock realised that the Earth's atmosphere is an unstable combination of elements, created and held in balance by life on the planet.

More people have taken an active interest in environmental issues

The idea came to Lovelock while working with NASA to uncover ways of identifying life on Mars. He came up with a radical suggestion: measure the chemical activity of the atmosphere.

This insight led Lovelock to the idea of the Earth as a single living organism, or self-regulating whole. According to the theory, the Earth maintains its climate and chemical composition at levels that are comfortable for living organisms. The theory was named Gaia, after the Greek goddess of the Earth.

Not all scientists found the 'Mother Earth' connotations to be palatable, and the theory of Gaia remains controversial.

However, thanks to Lovelock's theory a new kind of science was born, one that acknowledges the interaction between geology and biology – including human influences.

Also, many more people have taken an active interest in environmental issues, hoping to minimise the damage caused by humans to the planet.

SAVING THE PLANET

James Lovelock discovered the electron capture detector in 1957 after realising that 'nuisance' signals on another detector could become a powerful probe. But the device became one of the most powerful scientific tools for detecting harmful chemicals, revealing the extent of the damage they are causing to the atmosphere and Earth.

The findings established that chemicals are destroying the ozone layer

The device was a million times more sensitive than other detectors that were being used to monitor the environment at the time, boosting the efforts of scientists studying pollutants.

It was just what was needed to determine traces of pesticides in soils, food and water, when the effect of chemicals on many environmental systems was becoming an issue of public concern.

Within a year of its invention, the device was being applied to pesticide analysis. The harmful effects of pesticides, such as DDT, would have been very difficult to detect without the use of the ECD.

Using the device to take observations from the Shackleton research ship in Antarctica meanwhile, Lovelock demonstrated that chlorofluorocarbons (CFCs) were accumulating in the atmosphere without loss. The dramatic findings helped to establish that chemicals are destroying the protective ozone layer above the Earth.

Drummond 'Drum' Matthews and his research student Fred Vine made the discovery using a hand-hauled seabed 'magnetometer' during a University of Cambridge expedition to the Gulf of Aden.

The stripes, exactly mirrored on each side of the ridges, highlighted the reversals of the earth's magnetic field locked into the rock as it cools. The new sea floor had formed from up-swelling volcanic magma from beneath the Earth's surface.

The findings explained the huge (and previously mysterious) horizontal movements of land masses, observed from other evidence and predicted by the theories of continental drift and plate tectonics. Sea floor spreading confirmed the now widely accepted concept of floating continents.

The findings explained the huge horizontal movements of land masses

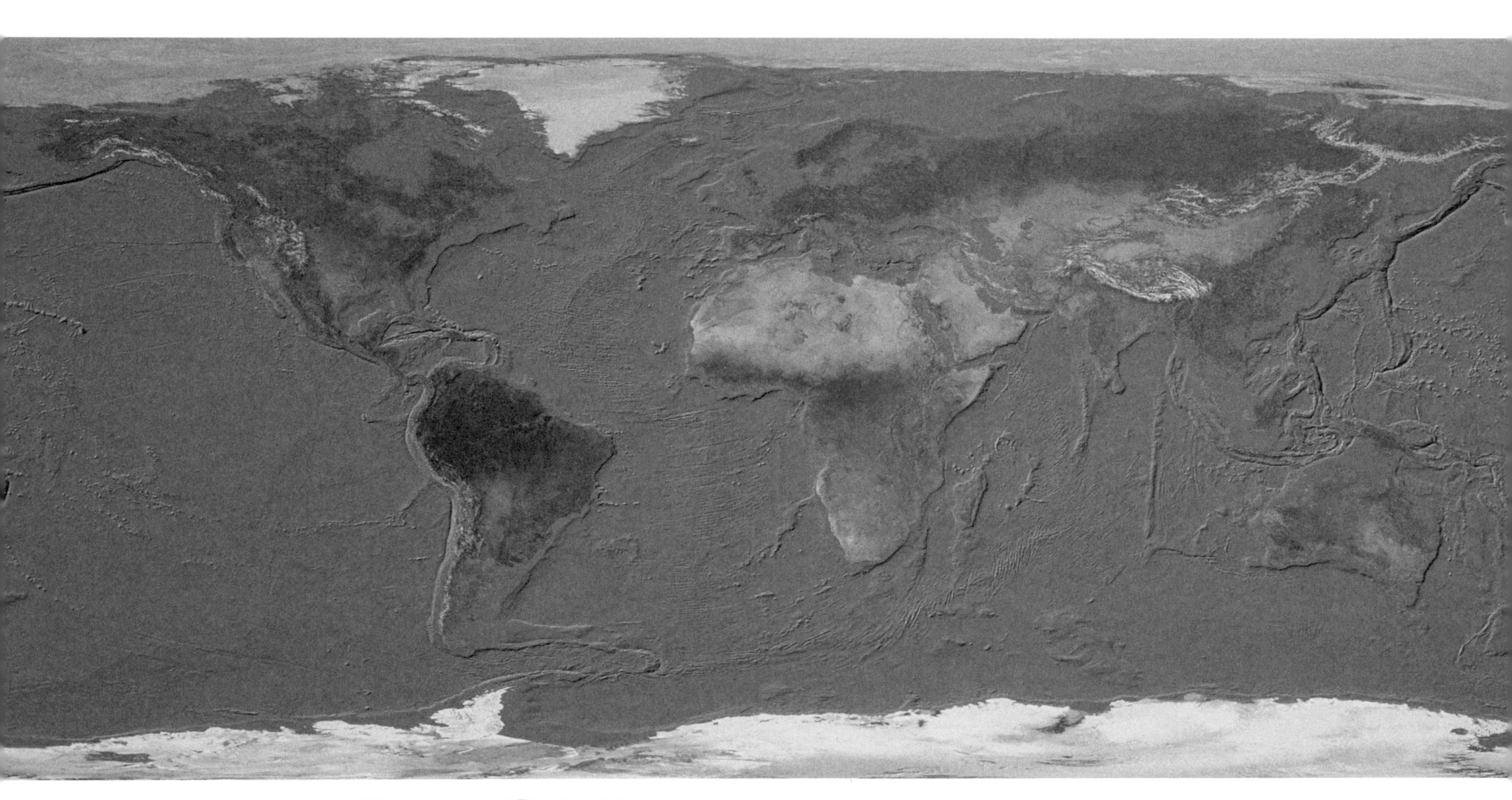

NEW GROUND

In 1963, two British marine geologists discovered huge matching magnetic 'stripes' in the rocks by ocean ridges. They had confirmed the controversial theory of sea floor spreading, a key precursor to the birth of plate tectonics.

THINGS ARE HEATING UP

In the 1950s, meteorologists assumed that the Earth's climate was more or less stable, and few people had then heard of 'global warming'. The pioneering climatologist Hubert Lamb was instrumental in establishing the study of climate change as a serious research subject.

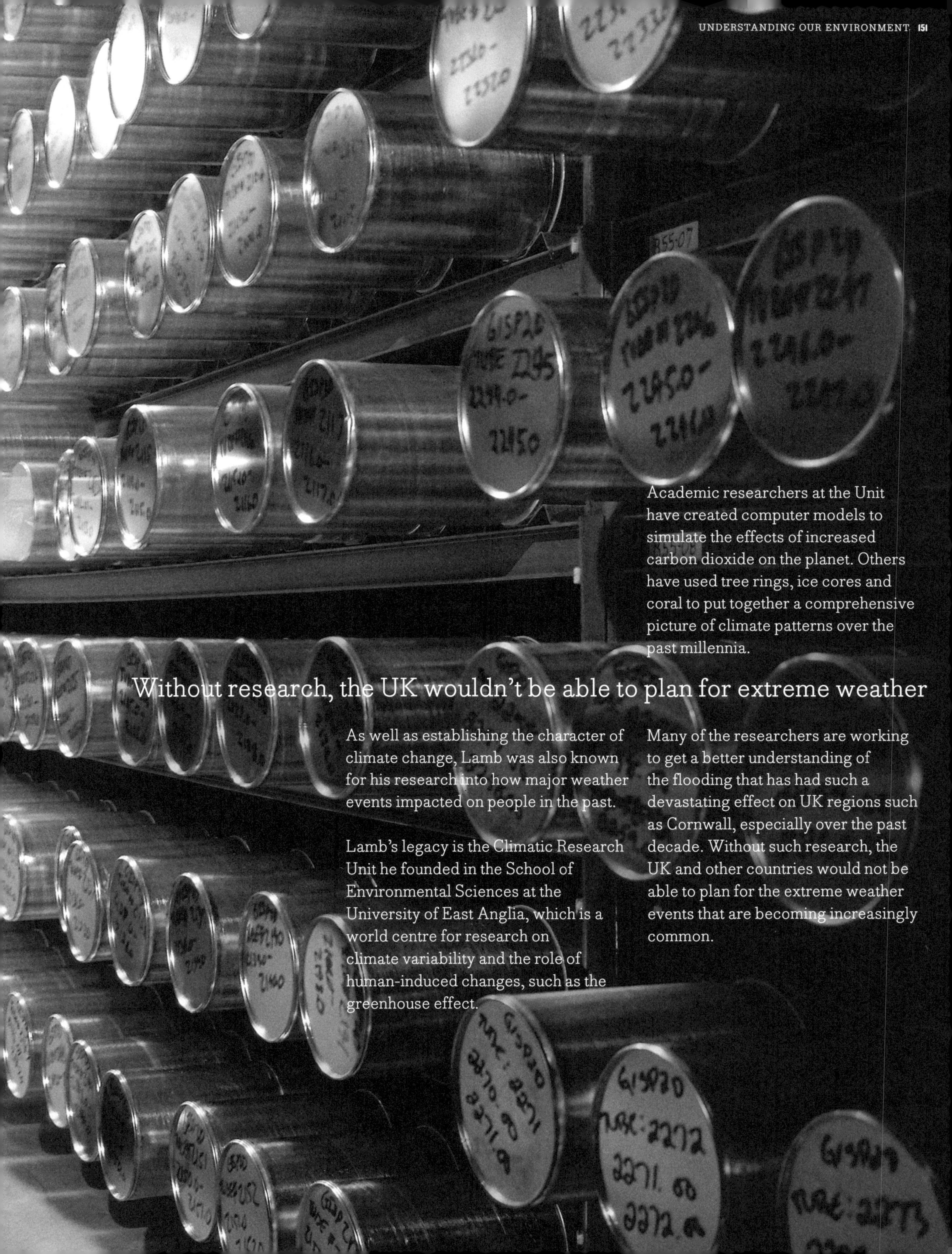

Academic researchers at the Unit have created computer models to simulate the effects of increased carbon dioxide on the planet. Others have used tree rings, ice cores and coral to put together a comprehensive picture of climate patterns over the past millennia.

Without research, the UK wouldn't be able to plan for extreme weather

As well as establishing the character of climate change, Lamb was also known for his research into how major weather events impacted on people in the past.

Lamb's legacy is the Climatic Research Unit he founded in the School of Environmental Sciences at the University of East Anglia, which is a world centre for research on climate variability and the role of human-induced changes, such as the greenhouse effect.

Many of the researchers are working to get a better understanding of the flooding that has had such a devastating effect on UK regions such as Cornwall, especially over the past decade. Without such research, the UK and other countries would not be able to plan for the extreme weather events that are becoming increasingly common.

LASTING EFFECTS

The instant catastrophic impact of flooding was brutally exposed by the tsunami disaster in late 2004, one of the world's worst natural disasters. But researchers at Middlesex University have shown that this is really just the start of the problems: people suffer from stress, sleeping problems and depression long after the flood-waters have receded.

The research has helped to raise awareness of these impacts in Government circles, increasing the emphasis on the 'social' consequences of flooding and helping to boost budgets for flood defence in the UK by hundreds of millions of pounds.

We have under-estimated the long term psychological impact of floods

Surveys of recent flood victims in England by the University's Flood Hazard Research Centre have found that we have up to now under-estimated the long term physical and psychological impact that flooding can cause.

The effects on individuals, households, and communities can last for years, with for example loss of possessions and the perceived security of the home causing lasting psychological damage, such as high anxiety and stress levels.

The lessons on how to assist different communities in dealing with flood emergencies and their aftermath are likely also to have continued value: flooding, locally and worldwide, is expected to increase over the next 50 to 100 years due to the effects of global warming.

JAMES LOVELOCK THE INVENTOR OF THE ECD AND THE GAIA HYPOTHESIS

Our atmosphere is an unstable combination of elements held in balance by the activities of life

"It was in some respects a lot easier when I was a young researcher. It is just in the nature of modern science. It is now more specialised and sub-divided; people know an awful lot about an awful little. My subject is geophysiology, but I was also working in the fields of chemistry and electronic physics, areas I was not qualified to work in. The secret of successful science is that it doesn't matter what you do, as long as you keep your eyes open.

We used to gather in the coffee room, where discussions were pretty spontaneous. People would ask: 'Jim, have you got an idea?' So it was with ECD (the electron-capture detector.) I invented the argon detector to help a colleague.

Working with a later detector, I noticed occasional odd signals. It was more of a nuisance at first. At the time I thought that the research would not have the slightest use to anybody, but it was good science. When I realised the signals emanated from elements, like halogens, it provided the perfect test for harmful chemicals and carcinogens.

I later applied for a grant to join the Shackleton research ship going to Antarctica. But the academic committee turned me down, believing that there could not be a device sensitive enough to measure chemicals in the atmosphere. I was however allowed to go on my own accord. It proved to be a very successful voyage. We demonstrated that chlorofluorocarbons (CFCs) were accumulating in the atmosphere without loss.

NASA was looking for ways to identify life on Mars. I told them that their methods would not be able to find life here, let alone on Mars. I was asked to come up with a better solution. By the following Friday I had come up with the answer – measure the chemical activity of the atmosphere. Our atmosphere is an unstable combination of elements held in balance by the activities of life. This would provide the undeniable signature of life.

In terms of acceptance, the Gaia hypothesis [the idea of Earth as a self-regulating living organism] has achieved something, but not in the delivery of actions. When it is realised that we are in for big trouble, then they will do something."

Lovelock received a PhD in medicine from the London School of Hygiene and Tropical Medicine in 1948 and a DSc in biophysics from the University of London in 1959. He was also based at the Medical Research Council's National Institute for Medical Research in London.

He is currently an Honorary Visiting Fellow of Green College, University of Oxford and has honorary Doctorates from the Universities of East Anglia, Plymouth, Exeter, Edinburgh, Kent and East London.

SPACE EXPLORATION

During the last 50 years UK astronomers, cosmologists and space technologists have answered some of the big questions about the universe, and produced technological spin-offs that have helped the inhabitants back home on Earth.

In the 1950s, Fred Hoyle explained part of the story of how we came to be formed: he showed how heavier elements, such as carbon and oxygen, were formed from hydrogen when stars exploded during the early stages of the Universe.

In the 1960s, a research student called Jocelyn Bell discovered (and named) pulsars – spinning neutron stars that eject streams of particles and light from their magnetic poles.

In the 1970s, a young mathematician called Stephen Hawking proved that singularities – infinitely dense points surrounded by black holes in space – were an inevitable consequence of Einstein's Theory of General Relativity.

During the same decade researchers at the University of Leicester observed the X-ray signals that suggested that black holes are common in the Universe. Their work has produced technological spin-offs completely unforeseen at the time: detectors developed for the NASA Chandra X-ray Observatory are now being used to monitor drug take-up in cancer cells.

Technological innovations have had far-reaching as well as more down to earth impacts. Martin Ryle created hugely powerful telescopes by arranging a number of smaller ones that work together – a method now used by astronomers all over the world.

Fred Taylor pioneered a technique called infrared remote sensing, which can monitor the heat radiated by the Earth's atmosphere, allowing us to predict the weather. IRS has been used on instruments on six Earth satellites and a series of planetary probes, to Venus, Mars, Jupiter and Saturn.

Scientists at the University of Surrey meanwhile designed low orbit satellites that have helped doctors and emergency workers in dangerous and remote regions of the world.

But then in 1957 Fred Hoyle (based at the Institute of Astronomy, at the University of Cambridge) and three fellow scientists proposed a startling theory; the elements were created in the oldest chemical factories in the Universe – stars.

Hoyle's theory showed how light elements changed into heavier ones

This theory – known as 'nucleosynthesis' – showed how light elements such as hydrogen were transformed into heavier ones, such as carbon and oxygen, during a star's death throes.

How are heavier elements created? Half a century ago this question still dogged scientists, who could not explain how elements heavier than hydrogen had formed in the early stages of the Universe.

ELEMENTAL PROCESS

The substances, which make up planets, moons, animals and people were born out of stars that exploded before our own Solar System even existed.

Elements are created through three distinct processes, with elements up to and including iron in the Periodic Table formed in a different way to heavier elements.

Hoyle later inadvertently coined the phrase 'Big Bang' when arguing against the theory that the Universe was created from a single explosion at the beginning of time. Big Bang theory today remains the accepted explanation for the creation of the Cosmos.

In 1965 postgraduate student Jocelyn Bell joined Anthony Hewish in the Astronomy Department of the University of Cambridge. At the outset, Hewish's idea was that his student should look for quasars, a rare cosmic object, that emits radio waves.

STAR STUDENT

Bell found evidence of a new type of star pulsating with radio waves

But Bell was to discover something far more surprising. The discovery would constitute a key breakthrough in our understanding of the chemical development of galaxies.

Over the next two years, the pair constructed a telescope as big as 60 tennis courts. Bell began to see patterns in the collected data early on.

One night in November 1967, using a faster data recorder, Bell's initial suspicions were confirmed: the radio source in the sky was blinking on and off every 1.338 seconds, like a radio beacon.

Over the next few weeks Bell found more of the pulsating radio wave sources in different parts of the sky. They provided the first evidence of a new type of star, one that pulsates with radio waves. Bell called it a pulsar.

We now know that a pulsar is a spinning neutron star. Jets of particles stream out of the top and bottom of the star from its two magnetic poles. As the star spins around, jets of particles produce powerful beams of light that sweep around in the sky like the light from a lighthouse.

From Earth, we see this light as radio waves. But because the beam rotates, the waves appears to blink on and off, like the view of a lighthouse spotlight in a storm.

DEATH STARS

Singularities – infinitely dense points in space with no dimensions where the laws of physics break down – must exist if Einstein's Theory of General Relativity is correct.

This mathematical fact was proved by Stephen Hawking as a graduate student at the University of Cambridge, working with the theoretical physicist, Roger Penrose at the University of Oxford in the 1960s.

They showed such phenomena were an inevitable consequence of Einstein's equations. When a large star runs out of its nuclear fuel and collapses, its matter crushes together at its centre forming a singularity. The surrounding region becomes a black hole, from which nothing can escape.

Hawking then applied quantum theory, the laws that govern subatomic particles, to black holes. He found that black holes must emit particles and radiation, as the particles dance over a point of no return known as the 'event horizon'. This is now called 'Hawking radiation'. Hawking also found that when holes finally die they explode with huge energy.

Hawking found when black holes die they explode with huge energy

In the 1970s Fred Taylor at the University of Oxford pioneered a technique that would be applied across the entire Solar System. It is called 'infrared remote sensing', and its power is the ability to monitor the heat radiated by the Earth's atmosphere, allowing us to predict the weather.

INFRARED FORECASTS

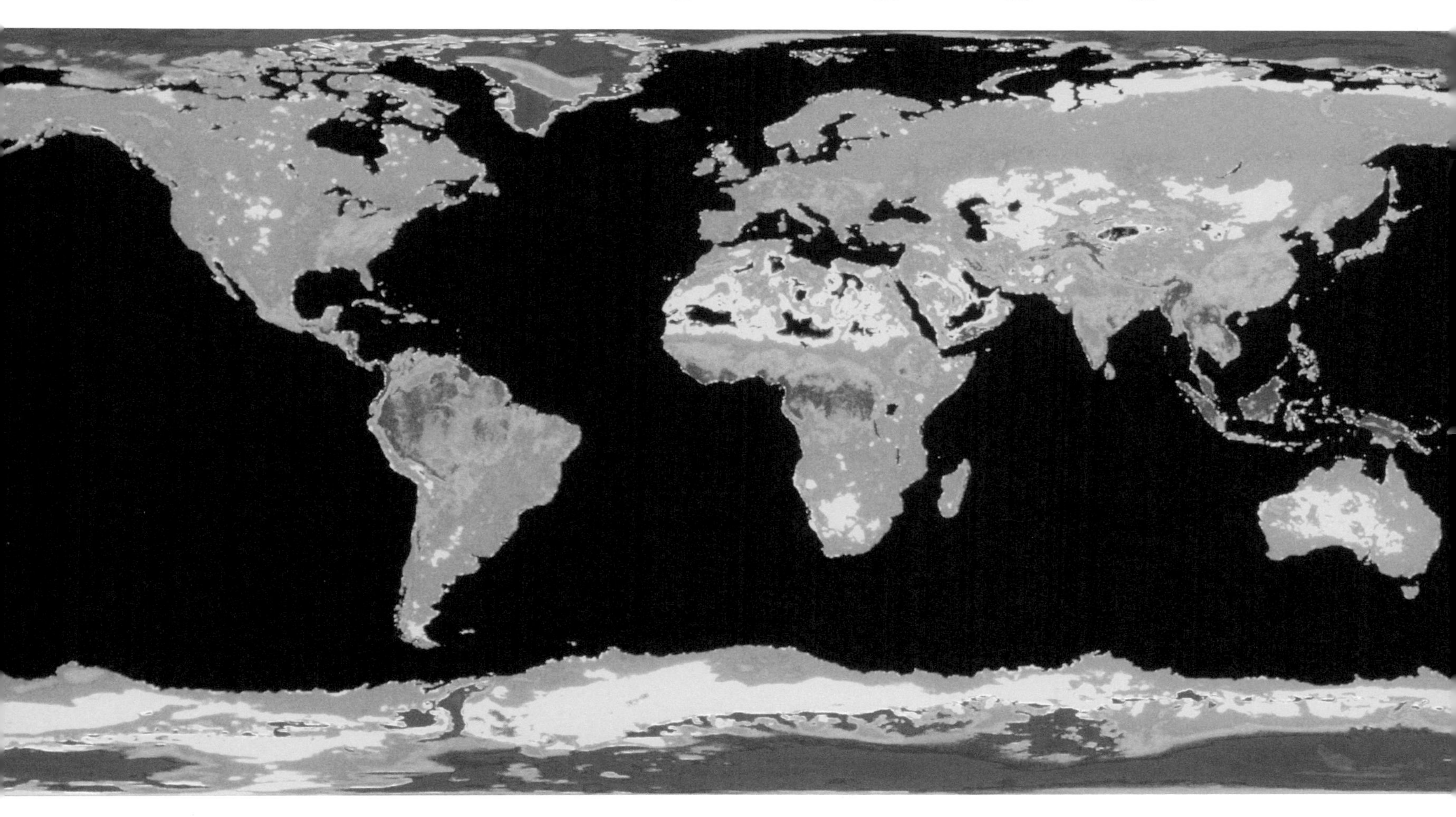

The method analyses heat radiation that the atmosphere emits in space

IRS has been used on instruments on six Earth satellites and a series of planetary probes, to Venus, Mars, Jupiter and Saturn.

The method analyses the heat radiation that the atmosphere emits in space, by means of spectroscopic instruments on an orbiting satellite. These look at spectral lines and bands in the emitted radiation to infer how much is present, plus the temperature and pressure. From this, humidity, and cloud and wind activity can be worked out and used to improve our understanding of the atmosphere, and to predict the weather and climate.

The approach has also contributed to efforts to understand the environmental hazards facing the planet. It has helped researchers investigating the physical processes underlying the greenhouse effect and its role in climate change, as well as the dynamics of the ozone layer which may lead to strategies to prevent holes in the ozone layer caused by pollution.

HOLE SPOTTING

The race to find black holes – regions in space where gravity is so strong that nothing, even light, can escape – has been on ever since they were postulated by Albert Einstein in his Theory of General Relativity.

Research by Ken Pounds and his team at the University of Leicester helped to provide the best evidence so far that black holes are common in the Universe.

In 1970s US astronomers found an intense X-ray source, later identified as a binary star system with one extremely heavy but small star, the characteristics of a 'stellar black hole'.

Using the UK's Ariel 5 satellite, the Leicester astronomers then observed X-ray signals thought to emanate from 'massive black holes' residing in the nucleus of active galaxies, the most powerful objects in the Universe.

In 1975 Ariel 5 monitored a powerful burst of X-rays from A0620-00, a new cosmic X-ray source subsequently understood in terms of a large amount of gas being dumped on to a black hole by the expansion of a nearby star.

A0620-00 provides a generic link between the first stellar mass black hole found in 1972 by US astronomers and the massive black holes later discovered by Ariel 5 lurking at the centre of many galaxies.

In 2001, the Leicester team uncovered the first evidence of an intermediate mass black hole in the nearby galaxy, Messier 82.

The University of Leicester provided evidence that black holes are common

In the early 1950s, scientists were busy looking at galaxies 1,000 million light years away and debating whether the Big Bang or Steady State theories described the nature of the Universe.

LOOKING BACK IN TIME

Like many scientists, Martin Ryle, an astronomer at the University of Cambridge, knew that the development of more powerful telescopes would hold the key to solving the puzzle. Looking at the light that has travelled to Earth from far away objects is like looking back in time, right to the beginning of the Universe.

By 1958, thanks to the development of a powerful computer built at Cambridge by Maurice Wilkes, Ryle could analyse data from many telescopes working within 5km of each other, just as if he had covered the whole area by a single vast device.

Ryle's approach has created devices that can see a postage stamp on the moon

In general, the bigger the telescope, the more powerful it is. So to see further into space – and hence further back in time – astronomers usually build bigger devices. But Ryle chose a different approach.

He created hugely powerful telescopes by carefully constructing and arranging a number of smaller ones that worked together.

Ryle's methods are now used all over the world. They even take advantage of the rotation of the earth to look at different parts of the sky. The approach, now known as the 'aperture synthesis technique', has created devices that are so powerful they can see a postage stamp on the moon. Observations made with these instruments have been crucial to the study of the stars and the study of the development of the Universe.

Despite the Internet, the world is far from a global village. In remote areas of the world where the nearest town or telephone is miles away, it is impossible to get news out, or vital expert advice in.

CRUCIAL LINKS

Since 1981 the team has been designing and constructing low-earth orbit satellites that are also relatively cheap – enabling more than just a few superpowers access to the latest space technology.

Low-cost satellites provide essential links for disaster relief worldwide

Such a situation can be perilous, particularly in the aftermath of natural disasters such as earthquakes or floods.

In these remote places, traditional satellites are too expensive for the communications that are needed. But thanks to Martin Sweeting and fellow academics at the University of Surrey, low-cost satellites now exist to provide the essential links for disaster relief all over the world.

Doctors working in remote regions in Africa and emergency personnel working to bring relief to flood and earthquake victims have all been helped by technology developed at Surrey.

KEN POUNDS THE MAN WHO FOUND BLACK HOLES IN THE UNIVERSE

The wider benefit, in my view, is much more in making science interesting to students

"My research field is X-ray astronomy, which effectively began in 1962 with the chance detection of a powerful cosmic X-ray source (Scorpius X-1) during a US rocket flight aimed at detecting solar X-ray fluorescence from the Moon. Riccardo Giacconi got the 2002 Nobel Prize for Physics for that discovery and follow-up. In the UK we were well placed to join the fun, having a competitive research rocket (Skylark) and access to the Southern Hemisphere sky from Woomera in Australia.

By 1970 some 30 cosmic X-ray sources were known, but their physical nature was largely unknown. The next milestone came in 1970 with the launch of the first X-ray astronomy satellite, Uhuru, again led by Giacconi's group. We were next into orbit with the UK's Ariel 5 satellite launched in 1974.

The best evidence for the existence of black holes comes from X-ray astronomy, and there we can share the credit for 'discovery' with the US. Uhuru established the first in our Galaxy by showing the powerful X-ray source Cygnus X-1 to be in a binary system where one stellar component was a normal star but the other a compact star too heavy to be a neutron star.

With Ariel 5, in turn, we were the first to demonstrate that Active Galaxies were generically powerful sources of X-rays, due to (in this case a supermassive) black hole. Our early success was due to being in the right place at the right time.

Application, in the wider sense, was never in our minds at that time. Applications however have arisen, more recently, from the technology developed for our space research. For example, the detectors we developed for the NASA Chandra X-ray Observatory are now being used to detect drug take-up in cancer cells via radioactive tracers. Another device will shortly be marketed as a lymph node detector.

What has also changed markedly since the 1970s is the willingness/pressure to seek such wider application. UK companies though are generally more risk averse.

Although there are significant examples of technology transfer from our 'search for black holes', the main wider benefit, in my view, is much more in making science – physics in particular – more interesting/attractive, especially to students.

Governments, Research Councils, Vice-Chancellors, will always wish to set priorities – and a minority may prove well chosen. However, history tells us a major part of public funding should go into 'blue skies' research."

NOBEL PRIZE WINNERS

Researchers based in UK research institutions winning the Nobel Prize since 1953

1953 (Physiology or Medicine)
Hans Krebs
'For his discovery of the citric acid cycle'

1954 (Physics)
Max Born
'For his fundamental research in quantum mechanics, especially for his statistical interpretation of the wave function'

1956 (Chemistry)
Cyril Hinshelwood
(Joint award with Nikolay Semenov)
'For their researches into the mechanism of chemical reactions'

1957 (Chemistry)
Alexander Todd
'For his work on nucleotides and nucleotide co-enzymes'

1958 (Chemistry)
Fred Sanger
'For his work on the structure of proteins, especially that of insulin'

1960 (Physiology or Medicine)
Peter Medawar
(Joint award with Frank Burnet)
'For discovery of acquired immunological tolerance'

1962 (Physiology or Medicine)
Francis Crick, James Watson, Maurice Wilkins
'For their discoveries concerning the molecular structure of nucleic acids and its significance for information transfer in living material'

1962 (Chemistry)
Max Perutz, John Kendrew
'For their studies of the structures of globular proteins'

1963 (Physiology or Medicine)
Alan Hodgkin, Andrew Huxley
(Joint award with John Eccles)
'For their discoveries concerning the ionic mechanisms involved in excitation and inhibition in the peripheral and central portions of the nerve cell membrane'

1964 (Chemistry)
Dorothy Hodgkin
'For her determinations by X-ray techniques of the structures of important biochemical substances'

1967 (Chemistry)
Ron Norrish, George Porter
(Joint award with Manfred Eigen)
'For their studies of extremely fast chemical reactions, effected by disturbing the equilibrium by means of very short pulses of energy'

1969 (Chemistry)
Derek Barton
(Joint award with Odd Hassel)
'For their contributions to the development of the concept of conformation and its application in chemistry'

1970 (Physiology or Medicine)
Bernhard Katz
(Joint award with Ulf von Euler, Julius Axelrod)
'For their discoveries concerning the humoral transmitters in the nerve terminals and the mechanism for their storage, release and inactivation'

1971 (Physics)
Dennis Gabor
'For his invention and development of the holographic method'

1972 (Physiology or Medicine)
Rodney Porter
(Joint award with Gerald Edelman)
'For their discoveries concerning the chemical structure of antibodies'

1972 (Economic Sciences)
John Hicks
(Joint award with Kenneth Arrow)
'For their pioneering contributions to general economic equilibrium theory and welfare theory'

1973 (Physics)
Brian Josephson
'For his theoretical predictions of the properties of a supercurrent through a tunnel barrier, in particular those phenomena which are generally known as the Josephson effects'

1973 (Physiology or Medicine)
Nikolaas Tinbergen
(Joint award with Karl von Frisch, Konrad Lorenz)
'For their discoveries concerning organisation and elicitation of individual and social behaviour patterns'

1973 (Chemistry)
Geoffrey Wilkinson
(Joint award with Ernst Fischer)
'For their pioneering work, performed independently, on the chemistry of the organometallic, so called sandwich compounds'

1974 (Physics)
Martin Ryle, Anthony Hewish
'For their pioneering research in radio astrophysics: Ryle for his observations and inventions, in particular of the aperture synthesis technique, and Hewish for his decisive role in the discovery of pulsars'

1974 (Economic Sciences)
Friedrich von Hayek
(Joint award with Gunnar Myrdal)
'For their pioneering work in the theory of money and economic fluctuations and for their penetrating analysis of the interdependence of economic, social and institutional phenomena'

1975 (Chemistry)
John Cornforth
'For his work on the stereochemistry of enzyme-catalyzed reactions'

1977 (Physics)
Nevill Mott
(Joint award with Philip Anderson, John H Van Vleck)
'For their fundamental theoretical investigations of the electronic structure of magnetic and disordered systems'

1977 (Economic Sciences)
James Meade
(Joint award with Bertil Ohlin)
'For their path breaking contribution to the theory of international trade and international capital movements'

1978 (Chemistry)
Peter Mitchell
'For his contribution to the understanding of biological energy transfer through the formulation of the chemiosmotic theory'

1979 (Physics)
Abdus Salam
(Joint award with Sheldon Glashow, Steven Weinberg)
'For their contributions to the theory of the unified weak and electromagnetic interaction between elementary particles, including inter alia the prediction of the weak neutral current'

1979 (Physiology or Medicine)
Godfrey Hounsfield
(Joint award with Allan Cormack)
'For the development of computer assisted tomography'

1980 (Chemistry)
Fred Sanger
(Joint award with Walter Gilbert)
'For their contributions concerning the determination of base sequences in nucleic acids'

1982 (Chemistry)
Aaron Klug
'For his development of crystallographic electron microscopy and his structural elucidation of biologically important nuclei acid-protein complexes'

1982 (Physiology or Medicine)
John Vane
(Joint award with Sune Bergstrøm, Bengt Samuelsson)
'For their discoveries concerning prostaglandins and related biologically active substances'

1984 (Physiology or Medicine)
Cøsar Milstein
(Joint award with Niels Jerne, Georges Køhler)
'For theories concerning the specificity in development and control of the immune system and the discovery of the principle for production of monoclonal antibodies'

1984 (Economic Sciences)
Richard Stone
'For having made fundamental contributions to the development of systems of national accounts and hence greatly improved the basis for empirical economic analysis'

1988 (Physiology or Medicine)
James Black
(Joint award with Gertrude Elion, George Hitchings)
'For their discoveries of important principles for drug treatment'

1996 (Economic Sciences)
James Mirrlees
(Joint award with William Vickery)
'For their fundamental contributions to the economic theory of incentives under asymmetric information'

1996 (Chemistry)
Harry Kroto
(Joint award with Robert Curl and Richard Smalley)
'For their discovery of fullerenes'

1997 (Chemistry)
John Walker
(Joint award with Paul Boyer)
'For their elucidation of the enzymatic mechanism underlying the synthesis of adenosine triphosphate (ATP)'

1998 (Economic Sciences)
Amartya Sen
'For his contributions to welfare economics'

2001 (Physiology or Medicine)
Paul Nurse, Tim Hunt
(Joint award with Leland Hartwell)
'For their discoveries of key regulators of the cell cycle'

2002 (Physiology or Medicine)
Sydney Brenner, John Sulston
(Joint award with Robert Horvitz)
'For their discoveries concerning genetic regulation of organ development and programmed cell death'

2003 (Physiology or Medicine)
Peter Mansfield
(Joint award with Paul Lauterbur)
'For their discoveries concerning magnetic resonance imaging'

pp. 6-7, Astronaut and satellite
NASA, Roger Ressmeyer, Corbis UK Ltd

HEALTHY BABIES AND BIRTH CONTROL
pp. 14-15, Contraceptive pills
Andrew Brookes, Corbis UK Ltd

p. 17, Louise Brown and parents
Adrian Arbib, Corbis UK Ltd

pp. 18-19, In-vitro fertilisation
Zephyr, Science Photo Library

p. 20, Sonogram scan
Rick Gomez, Corbis UK Ltd

p. 21, Newborns in nursery
Bsip, Beranger, Science Photo Library

pp. 22-23, Down's syndrome boy and brothers
Lauren Shear, Science Photo Library

p. 24, Hospital nursery
Ed Young, Corbis UK Ltd

p. 26, Edwards and Steptoe
Peter Price, Rex Features

HEALTHIER AND LONGER LIVES
pp. 28-29, Skeleton degenerating due to osteoporosis
Alfred Pasieka, Science Photo Library

pp. 30-31, MRI patient
Ed Eckstein, Corbis UK Ltd

pp. 32-33, Chemiluminescent liquid in spiral tube
Charles D Winters, Science Photo Library

p. 34, Surgeon using endoscope
Robert Llewellyn, Corbis UK Ltd

p. 35, Man with external pacemaker
Bettmann, Corbis UK Ltd

p. 36, Father and daughter brushing teeth
Tom Stewart, Corbis UK Ltd

p. 37, Thermograph of arthritic hip
Howard Sochurek, Corbis UK Ltd

p. 39, Portable defibrillator
Adam Hart-Davis, Science Photo Library

pp. 40-41, Needleless injection
Damien Lovegrove, Science Photo Library

pp. 42-43, Smoking in the '60s
Bettmann, Corbis UK Ltd

pp. 44-45, Vaccinating against Hepatitis B
Astier Frederik, Sygma, Corbis UK Ltd

pp. 46-47, Coffee plantation, Kenya
Allover Norway, Rex Features

p. 48, Sir Richard Doll
Nick Sinclair, Science Photo Library

MEDICINE UNDER THE MICROSCOPE
pp. 50-51, Model of a DNA Molecule
Digital Art, Corbis UK Ltd

p. 52, Protein structure
I Andersson, Oxford Molecular Biophysics Laboratory, Science Photo Library

p. 53, Insulin injection
Lester V Bergman, Corbis UK Ltd

pp. 54-55, Graphic of an endorphin
Jc Revy, Science Photo Library

p. 56, A test for antibodies
Lester V Bergman, Corbis UK Ltd

p. 57, DNA profiling
James King-Holmes, Science Photo Library

pp. 58-59, Dividing cancer cell
Science Photo Library

p. 60, Dolly the sheep
Jeremy Sutton Hibbert, Rex Features

pp. 62-63, Stem cell research
James King-holmes, Science Photo Library

p. 64, Rosalind Franklin
Science Photo Library

DISCOVERIES FOR THE DIGITAL AGE
pp. 66-67, Optical fibres
Lawrence Manning, Corbis UK Ltd

pp. 68-69, Drum scanner
Mark Gamba, Corbis UK Ltd

p. 71, Strip thermometer
Andrew Lambert Photography, Science Photo Library

p. 72, Man with holograms
Lawrence Manning, Corbis UK Ltd

p. 73, Atlas 1 computer
Sheila Terry, Rutherford Appleton Laboratory, Science Photo Library

p. 74, Scanning electron microscope
Colin Cuthbert, Science Photo Library

p. 75, Scanning in 3D
3D Scanners

pp. 76-77, Elderly woman
Ronnie Kaufman, Corbis UK Ltd

pp. 78-79, Solar panels in a 'solar farm'
Christopher J. Morris, Corbis UK Ltd

pp. 80-81, Nested fullerene molecules
Alfred Pasieka, Science Photo Library

p. 82, Dr Narinder Singh Kapany
Michael Rougier, Time Life Pictures, Getty Images

p. 82, Harold Horace Hopkins
University of Reading

PLANES, TRAINS AND AUTOMOBILES
p. 84, Land speed record attempt
Charles M Ommanney, Rex Features

p. 85, Directional signs
Richard Cummins, Corbis UK Ltd

pp. 86-87, Crater caused by a micrometeorite on the window of the space shuttle
NASA, Roger Ressmeyer, Corbis UK Ltd

pp. 88-89, Jet engine
Reuter Raymond, Sygma, Corbis UK Ltd

pp. 90-91, MagLev train
Sipa Press, Rex Features

pp. 92-93, 'Microcab'
Microcab Industries Limited

pp. 94-95, Trains at Glasgow Central Railway Station
Colin Garratt; Milepost 92 1/2, Corbis UK Ltd

pp. 96-97, Cabin simulator
Roger Ressmeyer, Corbis UK Ltd

pp. 98-99, London Underground
Touhig Sion, Sygma, Corbis UK Ltd

pp. 100-101, Railway viaduct arches
Robert Slade, Alamy

pp. 102-103, Police vehicle dealing with an incident
Shout

p. 104, Deborah Withington
Sound Alert Technology PLC

IDEAS FOR IDEALS
pp. 106-107, A hostel for the homeless
David Reed, Corbis UK Ltd

p. 108, Sorting grains
Pablo Corral V, Corbis UK Ltd

p. 109, Trading floor
Kai Pfaffenbach, Reuters, Corbis UK Ltd

pp. 110-111, Famine in Ethiopia
Patrick Barth, Rex Features

pp. 112-113, Child labour
John Stanmeyer, VII

p. 114, Schoolchildren
Bryn Colton, Assignments Photographers, Corbis UK Ltd

p. 115, Budget briefcase
David Levenson, Alamy

p. 116, Teacher reading to children
Guy Cali, Rex Features

p. 117, Elderly people waiting at bus stop
Helen King, Corbis UK Ltd

pp. 118-119, Sword and scales; the symbols of justice
Michael Nicholson, Corbis UK Ltd

p. 121, The Queen at the opening of the National Welsh Assembly
Graham Tim, Sygma, Corbis UK Ltd

pp. 122-123, Voters at a polling station
The Scotsman, Sygma, Corbis UK Ltd

p. 124, Chancellor Gordon Brown, General Election Campaign 2005 , Labour Party
PA, Empics

p. 125, Prime Minister Tony Blair, General Election Campaign 2005 , Labour Party
PA, Empics

p. 126, Sir Michael Rutter
Edwart Hurst, Rex Features

UNDERSTANDING OURSELVES
p. 129, Flying Fortresses in flight over Germany
Hulton, Deutsch Collection, Corbis UK Ltd.

p. 130, United Nations meeting
Bettmann, Corbis UK Ltd

p. 131, Pulling down the Berlin Wall
Sipa Press, Rex Features

p. 133, Tablet With Linear B Characters
Gianni Dagli Orti, Corbis UK Ltd

pp. 134-135, Celtic Torque
Werner Forman, Corbis UK Ltd

pp. 136-137, Summit of Industrialised Countries
Régis Bossu, Sygma, Corbis UK Ltd

pp. 138-139, Listening to the radio
Janet Wishnetsky, Corbis UK Ltd

p. 141, Welsh dictionary
University of Wales

pp. 142-143, Architecture of Adelaide House, London
Arcaid, Alamy

p. 144, Michael Ventris
Hulton Archive, Getty Images

UNDERSTANDING OUR ENVIRONMENT
p. 146, Satellite view of the Earth
Geospace, Science Photo Library

p. 148, Ozone hole over Antarctica
Rex Features

p. 149, The shifting continents
Tom van Sant, Geosphere Project, Santa Monica, Science Photo Library

pp. 150-151, Core samples from the Greenland ice sheet
Roger Ressmeyer, Corbis UK Ltd

pp. 152-153, Extreme weather
Michael Mulvey, Dallas Morning News, Sygma, Corbis UK Ltd

p. 154, James Lovelock
George W Wright, Corbis UK Ltd

SPACE EXPLORATION
pp. 156-157, Supernova explosion
Roger Harris, Science Photo Library

pp. 158-159, Pulsars
Celestial Image Co, Science Photo Library

p. 160, Black hole
Aaron Horowitz, Corbis UK Ltd

p. 161, Earth's climatic regions
NASA, Science Photo Library

pp. 162-163, Material spiralling into the black hole
Chris Butler, Science Photo Library

pp. 164-165, Telescopes
Roger Ressmeyer, Corbis UK Ltd

pp. 166-167, Satellite in low-earth orbit
NASA, Roger Ressmeyer, Corbis UK Ltd

p. 168, Ken Pounds
Ian Ridpath, Galaxy Picture Library